THE GREEN NEW DEAL

WHY WE NEED IT AND CAN'T LIVE WITHOUT IT
...AND NO, IT'S *NOT* SOCIALISM!

Larry Jordan

PAGE TURNER
BOOKS INTERNATIONAL

The Green New Deal
Why We Need It and Can't Live Without It
...And No, It's *Not* Socialism!

Copyright 2019 by Larry Jordan

All rights reserved under International and Pan-American Copyright Conventions. No part of this book may be reproduced or utilized in any form or by any means, electronic or mechanical, including photocopying, scanning, recording, or by any information storage or retrieval system, without permission in writing from the Publisher.

First printing 2019
ISBN 978-0-578-49360-2
www.pageturnerbooks.biz

ABOUT THE AUTHOR

Larry Neal Jordan began his journalistic career at age 11 and as an adolescent, he founded a newspaper and radio station—an interest in journalism that evolved into creating *Midwest Today* magazine nearly 30 years ago. For 22 years he has also produced a weekly radio show heard on 52 stations across the American Heartland. He has done one-on-one interviews with U.S. Presidents and Vice Presidents, plus scores of celebrities. An accomplished writer of long-form journalism, his scrupulously well-researched articles have been quoted in books on Abraham Lincoln, John Kennedy, Hillary Clinton, Billy Graham, Grant Wood and many others; and used as sources for documentaries and news reports by CBS, ABC and BBC-2. He has appeared on hour-long BBC radio broadcasts in the EU and public radio in America. Jordan's work has also been spotlighted favorably by various print and broadcast media worldwide.

AUTHOR ACKNOWLEDGEMENTS

I want to pay tribute to some early influences whose work fostered my appreciation for nature. Aside from my parents, Edward and Olga Jordan, I was most influenced by the writings of Gladys Taber (her *Stillmeadow* series of books); Hal Borland (author and *New York Times* editorial writer); and Marie Cornic (of Morning Sun, Iowa, a personal friend whose homespun columns celebrated the changing seasons). I was also inspired by the passion of John Denver for nature's grandeur, not only in the poetry of his songs but his hands-on work in the environmental movement. (I bumped into John while watching in the VIP area the NASA launch of the first woman on the Space Shuttle). So many others—Robert Redford, Carole King, et al—have brought public awareness to the threats we are facing to our planet's survival.

I am indebted to my beautiful daughter, Sara Jordan-Heintz—an honors graduate of the prestigious University of Iowa and winner in 2018 of the Genevieve Mauck Stoufer Outstanding Young Iowa Journalists Award from the Iowa Newspaper Association, plus a 2019 recipient of honors from the Associated Press. Sara authored the fun and fascinating 2019 book *Going Hollywood: Midwesterners in Movieland*. She provided invaluable assistance in scrutinizing my manuscript and providing input.

Table of Contents

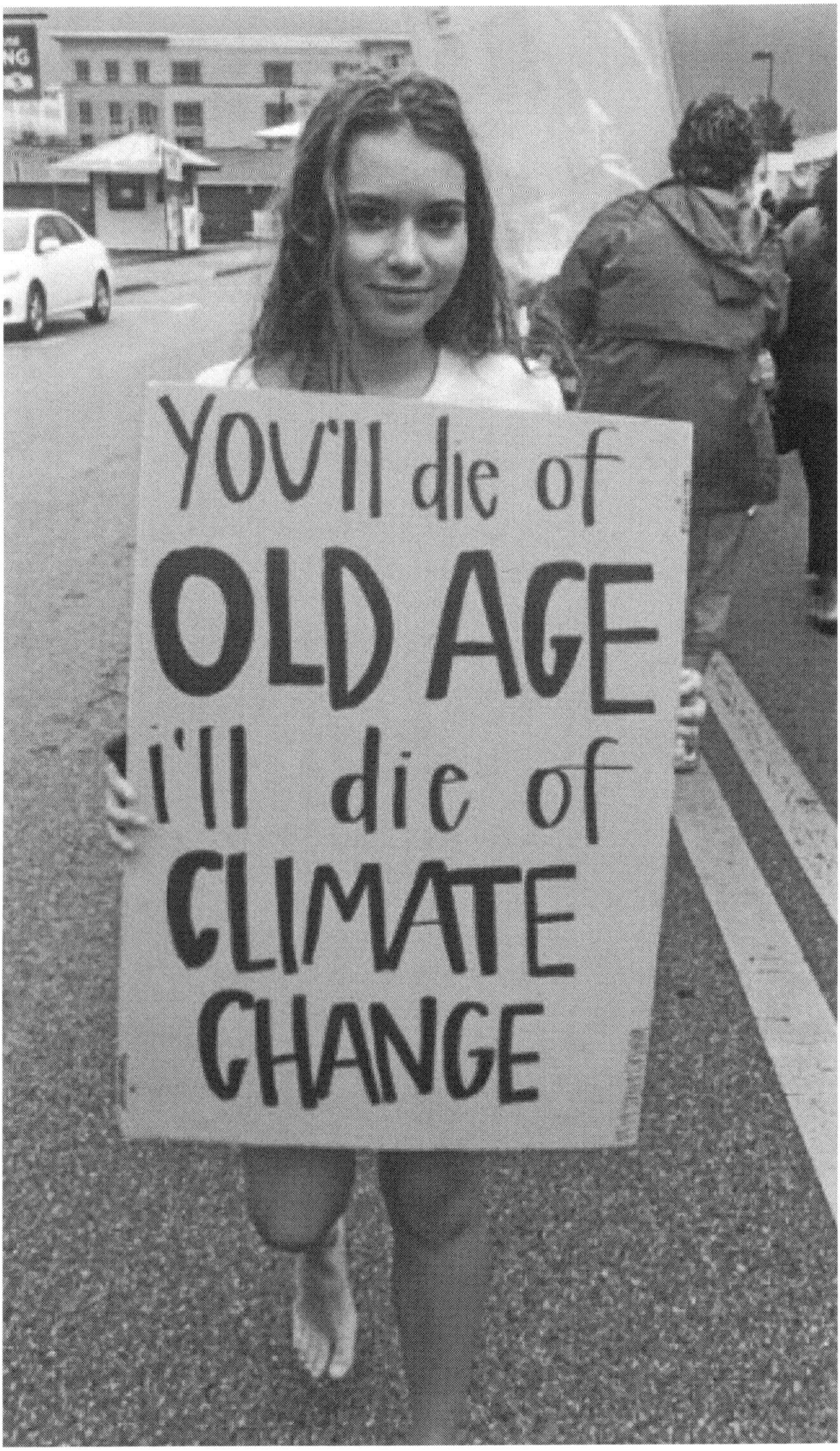
YOU'll die of
OLD AGE
i'll die of
CLIMATE
CHANGE

1.

What the Green New Deal is and Why It's Needed in the Trump Era

A NEW NATIONAL, SOCIAL, INDUSTRIAL AND ECONOMIC MOBILization on a scale not seen since World War II is underway, sparked by the realization by millions of Americans that unless our policies change, our warming climate, unbridled pollution, and economic injustice will tip the balance into catastrophic chaos for our nation and our world.

Multiple studies published in peer-reviewed scientific journals show that 97 percent or more of actively publishing climate scientists agree that climate-warming trends over the past century are extremely likely due to human activities. Nearly 200 worldwide scientific organizations agree.

A "Green New Deal" (GND) is being discussed that would address many of the threats that confront us, and is gathering momentum by the day. Simply put, the GND merely entails economic planning and industrial policy measures which would enable mobilization for the environment, similar to the economic mobilization for World War II, and not unlike the internal planning of large corporations.

The term "new deal" harkens back to the days of President Franklin D. Roosevelt and the compact he had with the American people to turn the ship of state around from the disastrous economic course on which it was heading. The nautical metaphor is apt, because FDR—a former assistant Secretary of the Navy—charted a new direction for the United States, with many innovative policies and bold actions that rescued us from great peril, ranging from our economic collapse caused by the Great Depression to the threat posed by a World War.

Over 400,000 Americans died in World War II, but the World Health Organization (WHO) estimates that the number of people killed globally today from climate change—warming-linked extreme weather, drought and disease—is at least 150,000 people per year. So isn't this an international emergency too?

Here are the goals of a Green New Deal

Broadly outlined, by the Sierra Club and other such groups, here are some of the primary goals for a Green New Deal that would help us swiftly transition to a clean energy economy:

•**Infrastructure Renewal:** We have a major, job-creating opportunity to repair, upgrade, and expand our country's neglected roads, bridges, energy grid, and water systems. This is not only a matter of fixing what's broken—it's a chance to build a cleaner, more affordable, and more resilient infrastructure system that supports workers and frontline communities for coming generations.

•**Climate sanity:** By investing in smart grids for renewable energy distribution, encouraging energy-efficient manufacturing, and expanding low-emissions public transit such as light rail, a Green New Deal would significantly reduce climate pollution.

•**Clean air and water:** A Green New Deal would replace lead pipes, clean up hazardous waste sites, and reduce toxic air and water pollution from oil, gas, and coal.

•**Community resilience:** Communities need greater resources to ensure safety and growth amid rising climate risks. A Green New Deal would help climate-exposed communities build bridges that can withstand floods, restore wetlands that buffer hurricanes, and shield coastlines from sea level rise.

•**Greater racial and economic equity:** A lion's share of the benefits of a GND would go to the working-class families and communities of color that have endured disproportionate economic and environmental hazards for decades. A Green New Deal must counteract systemic racism and economic exploitation by giving hard-hit communities priority access to new job opportunities, cost savings, pollution cleanup projects, and climate resilience initiatives.

•**Create family-sustaining jobs:** Each project should be required to pay workers prevailing wages, hire locally, offer training opportunities, and sign project labor agreements with unions.

•**Weatherize America:** Each time that a homeowner, business, or local government decides to weatherize a building, it supports jobs, slashes energy bills, and cuts climate pollution. A nationwide Green New Deal plan to weatherize buildings from coast to coast would create hundreds of thousands of retrofitting jobs, save families billions of dollars, and move us closer to climate sanity. We could achieve these goals with new national energy efficiency standards for public and private buildings, with public investments to help energy utilities implement the standards. New national standards for more energy-efficient appliances and industrial processes would create even more high-road jobs in manufacturing and engineering, while further cutting energy costs, toxic emissions, and climate pollution.

•**Buy Clean:** Each year the federal government spends billions of our tax dollars to buy goods, from steel for bridges to paper for offices. As part of a Green New Deal, a new "Buy Clean" law would ensure that these government purchases help fuel the transition to a clean energy economy and the creation of good jobs for those who need them most. Buy Clean standards would require, for example, that tax dollars be spent on goods manufactured with clean and efficient practices that protect our air, water, and climate.

As the Sierra Club reports, none of this is hypothetical. A lot of it is happening now. "From coast to coast, broad local coalitions are leading the way in pushing state-level Green New Deal policies that create good jobs, cut climate and local pollution, and counteract racial and economic inequity. As Donald Trump desperately tries to divide us, unions, environmental groups, and racial justice organizations are joining forces to chart the path for a Green New Deal. Their local successes offer momentum, and a model, for a nationwide mobilization under a new administration." Here are just a few examples:

•**Weatherization in Illinois:** One month after Trump's election, the Illinois Clean Jobs Coalition succeeded in getting the Future Energy Jobs Act signed into law, after two years of organizing and advocacy by unions, green groups, consumer associations, and environmental justice organizations. Among other things, the law sets new energy efficiency standards and invests in weatherizing buildings across the state. The gains for Illinois offer a glimpse of what a nationwide weatherization plan could offer: the creation of over 7,000 new jobs in the state each year, reduced air and climate pollution, and $4 billion in energy savings for Illinois families, with priority access for low-income households.

•**Buy Clean in California:** In 2017, California enacted a landmark Buy Clean law—the handiwork of a statewide coalition of labor and environmental allies. The law states that when California spends taxpayer dollars on steel, glass, and insulation for infrastructure projects, the state must prioritize companies that limit climate pollution throughout their supply chain. Thanks to the law, California will now leverage its spending to encourage climate-friendly manufacturing and local job creation—a sample of what a much larger, nationwide Buy Clean law could achieve.

• **Infrastructure Renewal in Pittsburgh:** The unions, community groups, and environmental organizations that make up Pittsburgh United's Clean Rivers Campaign have been pushing for job-creating green infrastructure projects that could drastically reduce flooding in some of Pittsburgh's vulnerable neighborhoods. They are one of many local coalitions across the country calling for, and often securing, public investments in green spaces to absorb rainwater, replacement of lead pipes, more resilient roads, and other critical infrastructure upgrades. Such fights help lay the groundwork for a national infrastructure renewal plan to simultaneously boost community resilience and create good jobs.

We're all in this together

The bottom line here is that—as President John F. Kennedy reminded us—"Our most basic common link is that we all inhabit this planet. We all breathe the same air. We all cherish our children's future. And we are all mortal."

Unfortunately, this movement for a Green New Deal is taking place in the context of America's domestic political upheaval, which centers around Donald J. Trump.

Like a disasterous President before him, George W. Bush, Trump lost the popular vote (with 2.9 million more votes cast for Hillary Clinton), but still took office due to the arcane, undemocratic and outdated Electoral College. A two-year investigation by the Special Prosecutor, Robert Mueller, established that Trump won in part because of the direct intervention and illegal help of a hostile foreign power—namely, Russia—which interfered in and undermined the fairness of our democratic election process. What possible motivation would Russia have to help Trump become President unless it was to gain strategic benefits, because of his buffoonery and/or the fact that he has received billions of dollars in loans from institutions like Deutsche bank—used by Russian oligarchs to launder money through real estate, including purchase of Trump condos? The Mueller probe also found extensive ties between members of the Trump campaign and Russian officials. Donald Trump has obliged Russian interests.

Trump has unmistakably exhibited a profoundly autocratic approach to governing by leading an all-out assault on our institutions, tradition of seeking consensus, and our nation's allies. He has pro-

Donald Trump's unprecedented use of inflammatory rhetoric has divided the nation and offended our longtime allies abroad. His environmental policies are alarming.

moted divisiveness through his vile and intemperate rantings, incessant posts on Twitter and personal attacks, obvious bigotry, and aggressive attempt to overturn decades of policies that have protected the public welfare. He even mimicked and made fun of a disabled man.

With the help of Republicans in Congress, Trump has championed legislation such as his massive tax cut for the rich and big corporations which have widened income disparity, undermined our nation's economy, and exploded the deficit. Compliant Republicans have changed the rules in the Senate so they can rubber-stamp even his most unqualified or corrupt nominees to the courts, as they seek to swing the bias of our judicial system to the extreme right and favor corporations over average citizens.

Trump has deliberately undermined the Affordable Care Act (ACA) to try to make it fail, probably because it is the signature achievement of our first black President, Barack Obama, and Trump has vowed to destroy every aspect of Obama's legacy. (Trump launched his current political career by promoting the "birther" conspiracy theory, claiming Obama wasn't born in the U.S., even though newspapers at the time published contemporaneous announcements of his birth and birth records prove he was born in Hawaii).

Even when Republicans in Congress failed to quash the ACA, Trump has persisted in his efforts to take life-saving health care coverage away from millions of Americans, totally indifferent to the human suffering this will cause.

You don't have to be a psychiatrist to know that the draft-dodging Donald Trump's extreme narcism is manifestly evident, along with his shocking ignorance, vindictiveness and seeming delight in embracing policies that harm people. Trump's habitual lying—which the *Washington Post* calculated as 9,014 "false or misleading claims" just during his first 773 days in office—has grown even worse of late, and reached a manic stage. Members of his team also lie constantly.

Using the formidable power of the "bully pulpit" and his ability to dominate every news cycle and claim the public's attention by his outlandish behavior, Trump has lead a frontal assault on a free press, dismissing as "fake news" any truths which he regards as inconvenient. This is the same strategy also pursued by some of the world's most despised dictators—Stalin, Hitler, and Mussolini. By labeling the mainstream media "the enemy of the people," Trump is literally using a phrase embedded in Article 58 of the Russian Criminal Code. Adopted in 1926, it gave that government broad power to arrest people and shut down media outlets. Trump has even ranted about parodies of him on the TV comedy series *Saturday Night Live* and warned of federal investigations—our First Amendment be damned. Trump has repeatedly vowed to change libel laws and "sue [media] like you never sued before." Russian-American author Anastasia Edel warns "it is not impossible to imagine a rise in predatory lawsuits against liberal media. And White House chief strategist Steve Bannon has summed up his own wishes for the press in a truly Leninist way, calling the media 'the opposition party' and advising it to 'keep its mouth shut.'"

Donald Trump has surrounded himself with acolytes and enablers within his administration while terminating those who might challenge his actions and assumptions, or refuse to carry out his illegal orders. He has given White House adviser Stephen Miller broad power to design a hard-line immigration enforcement strategy that courts have found to be illegal and inhumane. Ironically, Miller's xenophobia dates back to his high school days, despite the fact his own great-grandparents came to America as immigrants to escape persecution in Nazi Germany.

Trump has openly defied state, federal and international laws, allegedly promising to pardon those he has encouraged to break the law, and publicly attempted to intimidate or bribe witnesses in criminal proceedings.

As *Foreign Policy* magazine points out, "He has personally attacked judges for decisions he disagreed with and for their ethnic heritage, bemoaned the independence of the Federal Reserve, and directed the FBI and the Justice Department away from investigations aimed at his advisors and toward political opponents such as former Secretary of State Hillary Clinton. Trump's affinity for strongmen in Ankara, Manila, Moscow, and Riyadh, meanwhile, is well documented."

Following violence and torch-light parades in Charlottesville, Virginia—eerily reminiscent of the days of the racist KKK rallies—the President equivocated about condemning neo-Nazis, the alt-right, and white nationalists, despite their anti-American policies.

Trump is also destroying the U.S. government from within. When Mike Mulvaney, now acting White House chief of staff, took over the Consumer Financial Protection Bureau (CFPB)—the fledgling agency (which Elizabeth Warren helped create) that was supposed to protect consumers from predatory lenders and mortgage abuse—he

Torch-carrying Neo Nazis, alt-right, and white supremacists encircle and chant at counter protestors at the base of a statue of Thomas Jefferson at the University of Virginia in Charlottesville in 2017 before violence breaks out. Donald Trump stirred controversy by saying there were "very fine people on both sides."

de-constructed it with neglect and bureaucratic self-sabotage. This is a familiar Trump blueprint to neuter agencies that are supposed to protect Americans but end up doing the exact opposite. As a Republican former lawmaker, Mulvaney called the CFPB a "joke." He even endorsed abolishing it. When he took over the agency, he announced "regulation by enforcement is done. We're not doing it anymore." Mulvaney delayed the payday-loan rules, dropped lawsuits against payday lenders that victimize people (especially the poor), and stripped enforcement of fair-lending protections.

Yet inexplicably, Donald Trump has retained the unswerving loyalty of approximately 35 percent of the electorate. These Trump supplicants either believe every outrageous thing he says, or simply don't care that he is lying, despite his taking actions which hurt them. It is an odd phenomenon. They seem to be caught up in his larger-than-life cult of personality—not necessarily due to their own make-up, but because their belief system is fueled by propagandists like Fox News, Sean Hannity, Lou Dobbs, Laura Ingraham and Rush Limbaugh who distort and lie to promote Trump's image in a positive way.

Donald Trump publicly evinces symptoms which mental health experts describe as being sociopathic at the very least, and possibly pathological—dangerous traits to have in the leader of the free world.

Republican leaders privately disparage the President but have made a Faustian bargain and sold their souls to the devil because they know Trump will sign everything on their legislative wish list, especially more tax cuts for the rich. They are also afraid his rabid supporters will mount primary challenges to them in their re-election bids.

Trump's long-term connections to hostile foreign states, especially Russia, and parroting of their party lines while dismissing the concerns about America's national security raised unanimously by the U.S. intelligence community, represent an alarming departure from all of our nation's past Presidents, regardless of political party.

Trump's environmental damage is incalculable

Not surprisingly, our environment is also suffering at the hands of Mr. Trump. He withdrew our country from the Paris climate agreement. He has endorsed what Michael Flare has called "a wish list drawn up by the major fossil fuel companies."

It is estimated that due to the Adminstration's easing of federal pollution regulations on coal-burning power plants, there will be upwards of 1,400 extra premature deaths annually and 15,000 new cases of upper respiratory problems, a rise in bronchitis, and tens of thousands of missed school days. There will predictably be greatly increased levels of air pollutants like mercury, benzene and nitrogen oxides, thanks to its push to lift restrictions on air pollution. Trump has ordered the suppression of news about pollution science. He has made drastic cuts to the budget of the Environmental Protection Agency. He has opened America's waterways to far greater dumping of waste and pollutants, including mining waste. And this is just a partial list.

There are additional concerns about what has been happening to us—not just since Trump took office. Other Presidents and Congresses have also been neglectful in addressing pressing problems.

Jannat Shrestha enumerates the matter lucidly as follows: "The over consumption of resources and creation of plastics are creating a global crisis of waste disposal. Developed countries are notorious for producing an excessive amount of waste or garbage and dumping their waste in the oceans and, less developed countries. Nuclear waste disposal has tremendous health hazards associated with it. Plastic, fast food, packaging and cheap electronic wastes threaten the well being of humans. Waste disposal is one of the most urgent current environmental problems."

Human activity is leading to the extinction of species and habitats and the loss of bio-diversity. Ecosystems, which took eons to perfect, are in danger when any species population is disappearing. The balance of natural processes like pollination is crucial to the survival of the ecosystem and human activity threatens the same. The decline in the population of bees is one example. Another is the destruction of coral reefs in the oceans, which support the rich marine life.

Clean drinking water is becoming a rare commodity throughout the world. The availability of water, or lack thereof, is an economic and political issue as the human population fights for this resource. Industrial development and agricultural use of pesticides is filling our rivers, lakes and oceans with toxic pollutants which are a major threat to human health. Dirty water is the biggest health risk of the world. Run-off to rivers carries along toxins, chemicals and disease carrying organisms. Pollutants cause respiratory diseases like asthma and cardio-vascular problems. High temperatures encourage the spread of infectious diseases like Dengue fever in parts of the Third World.

Globally, 780 million people do not have access to a clean water source. An estimated 2.4 billion people lack adequate sanitation. Over 801,000 children younger than five years of age perish from diarrhea each year, mostly in developing countries. This amounts to 11 percent of the 7.6 million deaths of children under the age of five and means that about 2,200 kids are dying every day as a result of diarrheal diseases. Unsafe drinking water, inadequate availability of water for hygiene, and lack of access to sanitation together contribute to about 88 percent of deaths from diarrheal diseases.

According to the United Nations and UNICEF, one in five girls of primary-school age around the world are not in school, compared to one in six boys. One factor accounting for this difference is the lack of sanitation facilities for girls reaching puberty. Girls are also more likely to be responsible for collecting water for their family, making it difficult for them to attend school during school hours. The installation of toilets and latrines may enable school children, especially menstruating girls, to further their education by remaining in the school system.

We also have a problem with urban sprawl, which is the migration of people from high density urban areas to low density rural areas, which results in the spreading of a city over more and more rural land. This causes land degradation, increased traffic, environmental issues and health problems. The ever-growing demand for land also displaces important flora and fauna.

Genetic modification of our crops using biotechnology is called genetic engineering. This results in increased toxins and diseases as genes from an allergic plant can transfer to a target plant. Genetically modified crops can cause serious environmental problems as an engineered gene may prove toxic to wildlife. Another drawback is that increased use of toxins to make insect resistant plants can cause resultant organisms to become resistant to antibiotics.

Donald Trump is the best argument in favor of a Green New Deal

While the national trauma that we are all living through as a result of Donald Trump is so multi-faceted it's hard to focus on any one area, there is an increasing realization on the part of Americans across party lines that we urgently need a plan to tackle the twin crises of inequality and climate change. We are still a nation of vast public resources, but we need to marshal these to help transition us from an economy built on exploitation and fossil fuels to one driven by jobs that at least pay subsistence wages and assure a clean economy.

It simply does not suffice to maintain the status quo. Millions of citizens are being left behind, while the pockets of corporate polluters and billionaires are padded. Working class families of all ethnic backgrounds and colors are being victimized by stagnant wages, toxic pollution and dead-end jobs without the possibility of advancement.

Our worsening climate only exacerbates these systemic injustices, as we have become more vulnerable to extremes of weather—including bigger storms, longer droughts, worse winters and increased flooding.

The Green New Deal is not a cast-in-bronze plan as yet; it is still evolving. But there is an emerging consensus that we quickly need to upgrade our infrastructure, revitalize our energy system, retrofit our buildings as well as protect and restore our ecosystems. There is no reason why a sensible plan to accomplish these things could not also create millions of new jobs, with decent wages in the process.

❑

The purpose of this book is not to advance a specific proposal for a Green New Deal—since that is still in the formulation stage—but rather to survey what's been happening to our natural resources and our climate over the centuries.

To underscore the call to action, it's useful to examine the political context in which we are now operating; look at the history of North America and what steps various Presidents have taken on a bi-partisan basis to try to help safeguard our environmental heritage; assess our vanishing wilderness and how our National Parks are facing ruination due to decades of underfunding; discover how the federal government is harming what remains of our National Forests by allowing timber barons to exploit centuries-old trees for lumber; raise awareness about the peril our oceans now face; consider the unseen ramifications of Trump's plans to build a border wall; and learn the latest science on climate change.

This, then, is a *tour de force* of the problems which the Green New Deal must address—laid out in a simple, easy-to-understand way.

2.
The Political Context That's Inspiring A Green New Deal

J UST AS IN THE DAYS BACK IN THE 1930S WHEN PRESIDENT FRANK-lin Roosevelt proposed a New Deal for Americans, powerful and entrenched special interests are resisting the proposed changes embodied in today's GND, deliberately attempting in a concerted effort to misrepresent to the nation the true nature and cost of the course correction. Opponents are vigorously trying to mislead our citizens into believing the Green New Deal constitutes a covert attempt to transform our democratic republic into a socialist society, which many confuse with communism. But Americans' dismal understanding of our country's history prevents them from realizing that many of our Founding Fathers were socialists. They believed that "essential" services should be provided by government to the public at large for little or no remuneration. The costs of these services would be shared by the whole. This, by most modern accounts, is socialism.

Mark Brown of Capital University School of Law, comments that "The Constitution of the United States, drafted in the summer of 1787 in Philadelphia by some of the smartest men on this side of the Pond, proves this to be true. In that cherished document, the Found-ing Fathers demanded socialism. Section 8 of Article I, for example, empowers Congress 'To establish Post Offices and post Roads.' That same Section also authorizes Congress 'To raise and support Armies,' and even 'To provide and maintain a Navy.' Although the text does not preclude privatization of these public institutions—indeed, they continue to include entrepreneurial elements to this day—the Framers understood that they would certainly have public, social elements as well. Alexander Hamilton, James Madison, George Washington, Ben-jamin Franklin, and John Adams—among others—all signed this doc-ument. They agreed that the new national government would facilitate communication and defense through taxation. They agreed that these essential services would not have to be purchased on the open market. They agreed that these services would not be limited to those who could pay fair market value."

Brown points out that "The author of the Declaration of Independence, Thomas Jefferson (who skipped the Constitutional Convention in favor of traipsing off to Paris during that hot summer in 1787), also supported the fledgling nation's foray into socialism. Perhaps the greatest of all of America's socialized institutions, the nation's modern highway system, was begun in 1806 by then-President Jefferson's authorization of the Cumberland (National) Road. Transportation, too, was deemed to be one of the nation's essential services that could not be relegated to private industry.

"The Congress did President Jefferson one better. It socialized the great bulk of America's navigable waterways in the late 18th and early 19th centuries. The founding generation recognized early on that the national government needed the power to regulate interstate commerce—this was written into Article I of the 1787 Constitution—and waterways provided the most important channel of commerce. The national government, using this authority, opened America's internal waterways to commerce. These immense 'social' highways proved a boon to entrepreneurial activities (and perhaps saved the nation)."

Brown says that "Communication, transportation and mutual defense provide only the most obvious examples of the Founding Fathers' interests in socialized institutions.

"They were pragmatists, capitalists and socialists, willing to try whatever was necessary to insure that the American experiment did not fail. Of course, the Founding generation did not believe that every human endeavor benefited from governmental competition. The founding generation's socialism only went so far. The Founders believed in private enterprise.

"But," continues Mark Brown, "it was not long before the Founders' sons and daughters, grandsons and granddaughters, discovered the benefits of extending socialism beyond communication, transportation and national defense. Libraries, fire protection, police protection and education were all socialized to some extent in the 19th century. None of these developments replaced private enterprise— they merely insured that more Americans reaped the benefits."

Brown concludes that "History teaches us that the Framers were not averse to socialism. [But] they did not call it socialism. They called it good government."

The following two paragraphs warrant being in boldface type

to emphasize for readers of this book the fundamental strategy that animates the rightwing in America today:

The unspoken creed of Republicans is that the relative poverty of the masses has to be maintained to ensure their availability as wage slaves, to compete for access wages and produce surplus value for the profit of the minority owning elite. As Voltaire said, "The comfort of the rich is dependent upon an abundant supply of the poor."

The boogeyman of socialism repeatedly promoted in the public imagination by conservatives serves at least three main purposes: 1) to prevent people from realizing they are being taken advantage of; 2) to prevent the working class from coming together to make gains in the present society via unions, etc.; and 3) to protect profits and retain the dominance of the parasitical elites on whose behalf workers are exploited.

Viewed in this context, many of the otherwise confounding legislative initiatives, public policies and hypocritical lies perpetuated by Republicans who pretend to be doing the public good when actually doing the exact opposite, start to make sense from their own selfish strategic standpoint.

As Matthew Culbert of *Quora* opines, "The capitalist class compete against each other and go to war over trade...raw materials, markets and spheres of geopolitical advantage, but combine against the working class as one when their interests are threatened.

"Effectively the world's working class who comprise 90-95 percent of the world's population, presently run capitalism, from top to bottom, innovate, create, are scientists, teachers, administrators, [etc.].

"They just presently do not own and control the means and instruments for producing and distributing wealth"—and they are never going to unless they wake up and realize that the tactics of fear are being cynically used to confuse and paralyze them.

The same old strategy has been used for years

Paul Blumenthal, an alumnus of PBS's *Frontline*, points out in the *HuffPost* that "Every single political actor since the late 19th century advocating for some form of progressive social change—whether it be economic reform, challenging America's racial caste system or advocating for women's rights or LGBT rights—has been tarred as a socialist or a communist bent on destroying the American Free Enterprise System.

"Contemporary political conservatism has been focused on blocking social change that challenges existing hierarchies of class, race and sex since its founding in response to the French Revolution," Blumenthal observes. "Socialism emerged as the biggest threat to class hierarchies in due time and conservatives have called everything they don't like socialism ever since."

Blumenthal says "In the U.S., this dates back at least as far as the first Presidential candidacy of 'Prairie Populist' William Jennings Bryan in 1896. Bryan, who ran for President as a Democrat three times, was variously attacked as a 'socialist' and a 'communist' during each of his campaigns. His 1896 Presidential nomination took the Democratic Party on a 'perilous adventure in radicalism and centralization,' according to one conservative New York City paper. This despite the fact that he ran *against* actual Socialist Party nominees."

In the election of 1932, when Franklin Delano Roosevelt proposed reining in egregious abuses of corporate America, providing baseline economic security to impoverished citizens and empowering labor unions, Republican President Herbert Hoover attacked this New Deal as "a disguise for the totalitarian state." The American people were faced with a choice between "free enterprise" and "collectivism," he warned. Hoover lost the election in a landslide.

Paul Blumenthal says that "In what would become a very familiar line of attack, President Harry Truman's proposal for a national health care plan was labeled socialist as part of a public relations campaign run by the American Medical Association (AMA).

"The 1950s were marked by the rise of anti-communist red-baiters like Sen. Joseph McCarthy (R-Wis.) and the paranoid political movements that fueled his rise like the John Birch Society. McCarthy and the John Birch Society saw communists everywhere during the first years of the Cold War: Republican President Dwight Eisenhower's administration, including the Army, was filled with communists or maybe even the President himself was one."

When the Democrats proposed enacting a national health insurance program for the elderly, called Medicare, Ronald Reagan—then a Grade B movie actor with a rightwing mindset and political aspirations—attacked it as "socialized medicine." He said Social Security was "welfare" (but as President, later concocted a diabolical scheme with Alan Greenspan to embezzle trillions of dollars from Social Security to fund tax cuts for the rich and wage wars). Reagan also compared President John F. Kennedy to Karl Marx.

Segregationists claimed that the landmark Brown v. Board of Education decision, which banned school segregation, was the product of communist influence.

In 1988, Republican George H.W. Bush falsely portrayed his Democratic opponent, Governor Mike Dukakis as "far outside the mainstream."

Republican House Speaker Newt Gingrich labeled Bill Clinton's attempts at health care reform as "bureaucratic socialism." And when a bipartisan coalition in Congress enacted the S-CHIP program, which extended health coverage to children, the Heritage Foundation branded it as "a step towards socialism."

In 2008, as Democrat Barack Obama surged ahead of Republican Senator John McCain in the polls, the latter's supporters railed about "socialists taking over the country." Oil billionaire David Koch described Obama as "a hardcore socialist." Mitt Romney, who as Governor of Massachusetts had supported a plan of health care expansion in his state, nevertheless claimed that enactment of Obama's Affordable Care Act (ACA)—which gave over 23 million Americans health care insurance for the first time—meant that the U.S. had "effectively ceased to be a free-enterprise society." Really Mitt?

So Republicans have been trying to scare voters by warning about an imminent socialist takeover of the United States for about a century—and it hasn't happened yet.

Our democratic republic already encompasses elements of socialism, by the design of our Founding Fathers, and you don't hear many people complaining about paying taxes to fund public sanitation, roads, street lights, the police or fire departments, or other social benefits they receive.

But that hasn't stopped President Donald Trump and many present-day Republicans from seizing upon the Green New Deal to try to frighten people and thereby preserve the status quo. They are simply serving their paymasters—the fossil fuel industry and big corporations demanding a rollback of environmental regulations which have been enacted on a bipartisan basis over the decades.

The Grand Old Party is singing from the same old songbook

Writing in *The New York Times*, Lisa Friedman notes that "President Trump has claimed the Green New Deal will take away your 'airplane rights.' Senator Tom Cotton, Republican of Arkansas, told Hugh Hewitt, the conservative radio host, that the proposal would 'confiscate cars' and require Americans to 'ride around on high-speed light rail, supposedly powered by unicorn tears.' And Senator John Barrasso, Republican of Wyoming and chairman of the Committee on Environment and Public Works, warned that ice cream, cheeseburgers and milkshakes would be a thing of the past because under the Green New Deal, 'livestock will be banned.'"

Ms. Friedman states unequivocally: *"The [Green New Deal] doesn't do any of those things."*

Sadly, the Grand Old Party (GOP)—which used to have a proud record of opposing slavery, supporting a strong national defense, promoting fiscal responsibility, preserving our natural heritage and protecting individual rights—has abandoned all pretense of civility or rationality in public discourse and been transformed into a Trump-led cult disconnected from its roots. They think the more inflammatory and outrageous the rhetoric, the better to divide and conquer Americans. They don't even oust the self-avowed white nationalist Rep. Steven King (R-Iowa) from their ranks.

So we have the specter today of most Republican politicians shamefully adopting the practice of distorting the facts and outright lying during the course of public discussion of the issues. The party has employed the political equivalent of the magician's "slight of hand" or misdirection: they try to distract you while they're picking your pocket. They use scare tactics; they want you to be afraid, *very* afraid!

They try to convince you that it's not big corporations and the super rich and their outrageous tax breaks that are bankrupting our country, it's poverty-stricken Americans, many of whom are working two or even three jobs for wages that still don't cover the basic necessities of life. There is no state in the union in which the minimum wage suffices to support subsistence living, but the GOP opposes efforts to increase the minimum wage. What are people to do?

Demagoguery is epidemic in today's Republican party across the board, on issue after issue—and citizens need to be aware of these tactics when evaluating the merits of the current debate over the Green New Deal.

No one embodies this putrid mindset better than current Senate Majority Leader Mitch McConnell. Over the years, he has single-handedly blocked some of the most important legislation that would help Americans, by refusing to even schedule it for a vote. He is a vile and hypocritical obstructionist. He wouldn't even meet with President Obama's nominee for the U.S. Supreme Court, Judge Merrick Garland, despite the Constitutional right and duty that Obama had to fill the vacancy on the court which occurred as a result of the death of Antonin Scalia. McConnell stonewalled for a year, until Donald Trump took office, and then hamstrung Democrats who wanted to adequately investigate serious allegations of past sexual misconduct by Trump's nominee, Brett Kavanaugh—a Republican who ironically had been the lead author of the Ken Starr report that detailed Bill Clinton's alleged sexual misconduct in pornographic detail and was also involved in the effort to stop the vote counting in Florida which led to George W. Bush taking office even though Al Gore won more votes in 2000.

Recently, Mitch McConnell vowed to be the "Grim Reaper" who will kill what he labeled "socialist leg-islation," including the Green New Deal. Speaking to a group of his supporters in Kentucky who applauded him wildly, McConnell said "I don't want you to think this is just a couple of nutcases run-ning around on the fringe. This is perva-sive policy view on the other side... Are we going to turn this into a socialist country? Don't assume it cannot happen," he said. "If I'm still the majority leader of the Sen-ate, think of me as the Grim Reaper. None of that stuff is going to pass. None of it."

Meanwhile, a firm tied to a Russ-ian Oligarch that the United States Con-gress—both Republicans and Demo-crats—voted to impose sanctions on (but

Senator Mitch McConnell

which Trump lifted), is potentially pouring millions of dollars into Mitch McConnell's state of Kentucky to help build a new aluminum plant there.

The "Fox Factor"

Tune in Fox these days and you'll see a steady parade of so-called experts denouncing the Green New Deal and frenetically spreading outrageous lies to confuse viewers about the costs and polit-ical implications of the GND.

It undermines our democracy when millions of Americans rely on sources like Fox News (a.k.a. Faux News). The network is owned by Rupert Murdoch, a far-right Australian-born media baron who de-livers TV content to five continents, dominating Britain, Italy, and wide swaths of the Middle east. He publishes 175 newspapers, including the *New York Post* and the *Times of London*. In the U.S. alone he owns 45 percent of television stations in the nation—something Reagan's lifting of regulations over the broadcast industry helped facilitate.

Murdoch's media empire has enabled the rise of wacko, fringe candidates who—in an earlier time—wouldn't have a snowball's chance in hell of ever even being taken seriously, let alone elected to public office. Fox promoted the so-called Tea Party as a grassroots movement when it was largely funded by billionaire oil barons like the Koch Brothers. Murdoch plays his audience for suckers.

As *The Guardian* newspaper in England reports about Rupert Murdoch, "The media mogul spent decades coarsening the culture of news with his mix of snarling politics, moral outrage and sexual titillation. In shaking up tabloid newspapers in Britain and news broadcasting in America, he may claim to have toppled an establishment. But his brand of vituperative journalism also contributed to the rise of Brexit and Donald Trump—whom Mr Murdoch continues to support—even if his tabloid invective has been overtaken by online hard-right propagandists."

Fox News is the brainchild of the late Roger Ailes, the Republican dirty trickster who ultimately was ousted from the network after sexual harassment scandals and multi-million dollar settlements with his alleged female victims became public. Ailes was less a believer in family values than a hustler and an opportunist, who crafted Fox to become, in the words of veteran CBS newsman Dan Rather, "the closest we have come to having state-run media; a straight-up propaganda outlet... By any objective analysis, this is by far the closest we've come to Radio Moscow."

Indeed, Trump pays obeisance to Fox, and there is a symbiotic relationship between the two. The President has even been hiring some of the network's staff members to work in his administration, despite their lack of qualifications and unfitness to hold public office.

So political discourse in America has been sullied by the proliferation of outright lying—not just by the President of the United States, but by members of his party. Coupled with the outrageous lying by some of the major media like Fox, Americans are less well-informed than their counterparts in other countries.

Letting our personal biases determine the information we receive is a recipe for disaster. As Ray Dalio writes in *Principles*: "There is nothing more important than understanding how reality works and how to deal with it."

Barry Ritholtz reflects, "As a country, not all of us are working with objective facts. Opinions, wishful thinking, desire, hope, self-interest, and partisan politics, yes, but objective reality as it exists outside of ourselves, *not* so much. This raises problems for us as a functioning nation. When it comes to misinforming the American public about the issues that matter to policymakers, many are quick to blame the internet. [But] Fox News plays by rules very different from the rest of the media. They are not constrained by facts; make few attempts at being empirical or data-based; intellectually, they affiliate themselves not so much with a school of political thought as a specific political party. The rest of the media might be educated and urbane and therefore be left of center on social issues, but that is an ideological bias, not a partisan one. It's almost as if 'Fair and Balanced,' which Fox News used as a slogan, was purposefully sarcastic."

Ritholtz asks "If informed voters cast votes in their own best interest for the candidate they believe is most qualified, what does it imply in a democracy when huge swaths of the populace are uninformed? Once the population of a country no longer understands the world around them—or even the basic problems in their own neighborhoods—can they still govern themselves? Eventually, elections begin to resemble something looking like random outcomes. This lack of rationality eventually handcuffs policymakers, leading to poor societal outcomes."

Scapegoating immigrants to appeal to a bigoted base

Donald Trump, who—along with his late father, Fred—has a long history of racial discrimination, even resulting in lawsuits and huge fines, continues to scapegoat minorities. Central American immigrants fleeing violence in their home countries—which have been devastated by climate change and U.S. foreign policy over the last hundred years—are repeatedly referred to derisively by Trump as a "caravan" of "invaders" and "cold stone criminals" and "rapists and murderers" who will "infest" our country. He has grossly exaggerated how many have approached our southern border with Mexico and knowingly mischaracterized them, despite the clear evidence they consist disproportionately of poverty-stricken women and children, most of whom are walking—often barefoot—over a thousand miles trying to reach America. Under U.S. laws, crossing our border is *not* a felony.

Most families seeking asylum are not criminals—many are fleeing the Northern Triangle of Central America (El Salvador, Honduras and Guatemala), one of the deadliest regions in the world, where their communities are overrun by organized crime, violence, human trafficking and persecution; plus the direct effects of climate change that have spawned weather extremes that have decimated food supplies and local economies. They seek refuge and compassion from our nation, an offer of sanctuary emblematic of the statue of Lady Liberty that has welcomed immigrants for many years. These children and their parents have suffered irreparable harm and trauma.

The common rejoinder one hears is that these people should at least enter the United States legally or simply "get in line" if they are unauthorized. These suggestions miss the point: There is no line available for unauthorized immigrants and the "regular channels" do not include them.

Most people fleeing their home countries cannot access humanitarian protection. To be admitted as refugees, individuals must be able to prove that they have a "well-founded fear of persecution based on race, religion, membership in a particular social group, political opinion, or national origin." Asylum seekers are individuals already in the United States who fear returning to their home countries, and they must prove they meet the definition of a refugee. An immigrant does not qualify as a refugee or an asylee because of poverty or difficult economic conditions in their home country. There are limited forms of temporary humanitarian protection available, but these are rare.

Even before Trump announced "extreme vetting," it took four to eight years for most immigrants to enter the U.S. legally, due to all the red tape. Even those who can get in line are subject to long backlogs and waits. *Yes, we have a border crisis, but our leaders need to solve it.*

As President Donald Trump rails against illegal immigration, his administration has taken measures to restrict the ability of recently graduated and skilled foreign nationals to work legally in the U.S. As a result, many students and skilled immigrants are heading to Canada, where it's much easier to stay and work, according to immigration lawyers.

In fairness, it must be noted that under President Barack Obama, the U.S. set a record for deportations. The number of people fingerprinted, processed, and officially deported to their home countries soared under Obama, with 2013 setting a high-water mark of 435,498. *New York* magazine relates that "Responding to a 2014 crisis involving thousands of unaccompanied Central American minors attempting to cross the border, the Obama administration responded by detaining thousands of immigrant families and paying the Mexican government to intercept families before they reached the U.S., actions that drew sharp—if not widespread—criticism for treating the crisis more as a border-security issue than a humanitarian one. Two years later, some families who made the journey during that time and settled in the U.S. were deported."

The magazine reports that "The Obama administration also embraced, and defended, the use of privately run family detention centers, the unpleasantness of which seemed to serve as a deterrent for future Central American immigrants." So as Trump deports thousands, "it is with the help of an enforcement machine Obama helped construct, or at least maintained."

Still, it must be acknowledged that, as *New York* points out, "The Obama administration did not intentionally target longtime residents of the United States for removal; it did not separate parents from their children to extract concessions from the opposition party; it did not attempt to crack down on legal immigration to satisfy its base; it was not, in short, guided by the revanchist nativism that rules the day in the Trump White House."

The Trump administration and former Attorney General Jeff Sessions adopted unAmerican policies to unlawfully and permanently separate even babies and young children from their immigrant parents at the border. Kids have been locked in cages and detention centers by U.S. Customs and Border Protection (CBP) indefinitely in deplorable conditions, and later transferred to group shelters and foster care settings by the Office of Refugee Resettlement (ORR) in the Department of Human Services (DHS). Officials have consistently relied on deceitful practices to separate children from their parents, such as telling parents that their kids are being taken for a bath, when in reality they are being whisked away, never to be seen again.

After a massive, bipartisan public and Congressional backlash against his family separation policy at the U.S. border with Mexico, Donald Trump did what he usually does: he lied, and falsely blamed Democrats, claiming there is a law requiring it.

The Trump policy of forcibly breaking families apart at the border, and the refusal to grant asylum to refugees, violates the letter and the spirit of numerous U.N. treaties and protocols, including the "Convention Against Torture," "the Right to Family Life," "the Universal Declaration of Human Rights" and the "Recommended Principles and Guidelines on Human Rights At International Borders."

Federal courts have found it inexcusable and illegal for the Trump administration to be deporting the parents of these children without due process or adequate legal representation, and without a system in place for even tracking these youngsters or ever reuniting them with their families. The trauma and abuse the Trump administration has sadistically inflicted on innocent children are literally crimes against humanity and are punishable under international laws.

(The first prosecution for crimes against humanity took place at the Nuremberg trials. Unlike war crimes, crimes against humanity can be committed during peace or war. They are not isolated or sporadic events, but are part either of a government policy or of a wide practice of atrocities tolerated or condoned by a government or a *de facto* authority. Dehumanization, deportations, extrajudicial punishments, kidnappings and forced disappearances, military use of children, unjust imprisonment, political repression, racial discrimination, religious persecution and other human rights abuses may reach the threshold of crimes against humanity if they are part of a widespread or systematic practice. That definition applies to the Trump administration's border policy.)

The American Federation of Teachers (AFT) has fought back against this inhumane treatment of children and even presented a formal complaint for action to the United Nations Human Rights Council. The teachers point out that "denying the nurturing care that babies, toddlers and young school-age children require for proper development may cause these children in detention to suffer irreversible neurological and developmental damage."

The Trump adminstration continues to defy court orders to stop these unconscionable practices, and it's only growing worse. There are so many scandals swirling around Trump, the media and the American public can't focus on all the atrocities that are occurring.

Says *The Intercept*: "When Trump appoints Steve Bannon and Sebastian Gorka and Stephen Miller to his administration, and lets them decide immigration policy, that's white nationalism. When he talks about migrant caravans and gangs invading and infesting the United States, he's borrowing language and imagery straight out of the white nationalist movement, and specifically a racist French novel called the *Camp of the Saints*. When he tells his aides that he wants fewer immigrants from 'shithole' countries in Africa and more immigrants from places like Norway, that's white nationalism."

Even Fox News anchor Shep Smith debunked Trump's claim, saying "there's no invasion."

Nevertheless, President Trump defied Congress and overruled his military advisers and melodramatically ordered U.S. troops to amass along our southern border with Mexico, to appeal to his bigoted and/or uninformed base, and try to distract the country from the multitude of scandals enveloping his administration.

What all this debate about border security overlooks is that United States foreign policy created a lot of the problems which now afflict Central America and its people (*which will be discussed more thoroughly in Chapter 11*).

Climate change linked to the refugee crisis

The reason Donald Trump's border policies are relevant to a book on the Green New Deal is that the flight of refugees is precipitated in many ways by the effects of climate change on the economies of their home countries. So there is a direct correlation between the two.

The *Texas Observer* reports that "In the poor, violent Northern Triangle of El Salvador, Honduras and Guatemala, worsening floods, drought and storms are pushing a growing number of migrants north."

Gus Bova of the *Observer* explains that "A narrow strip of land flanked by oceans, Central America is one of the world's most environmentally vulnerable regions."

María Cristina García, a Cornell University professor of American studies who's writing a book about climate refugees, confirms "It's an area hit by hurricanes on both sides, rocked by volcanic eruptions, drought, earthquakes, and with accelerating climate change, it's even more vulnerable."

Bova says that "Central America hosts both spectacular catastrophes—Hurricane Mitch displaced three million people in 1998—and slow-burn disasters, such as frequent droughts worsened by climate change. Many of the current crop of refugees hail from the region's 'dry corridor,' a zone afflicted by alternating drought and flooding, where farmers face crop failure even without the effects of a warming planet. The corridor falls mostly within the poor, violent Northern Triangle of El Salvador, Honduras and Guatemala—a major source of immigrants to the United States."

Axios says that "the U.S. military views climate change as a threat multiplier, one that is likely to worsen already existing weaknesses of government and poverty."

In 2017 alone, nearly 19 million new internal displacements were recorded in more than 130 countries worldwide, largely triggered by extreme weather events such as floods and tropical cyclones, according to the Internal Displacement Monitoring Center. That's more than were displaced due to armed conflict in the same year, and climate change is aggravating many of these extreme weather events.

Of the 20 million refugees thought to exist worldwide today, only one percent (200,000) are permanently resettled. Many are staying put in camps, hoping things stabilize in their home countries.

Jeffrey Sachs, a professor and director of the Center for Sustainable Development at Columbia University, asserts that "President Donald Trump, former Florida Governor (and now U.S. Senator) Rick Scott, Florida Sen. Marco Rubio, and others who oppose action to address human-induced climate change should be held accountable for climate crimes against humanity. They are the authors and agents of systematic policies that deny basic human rights to their own citizens and people around the world, including the rights to life, health, and property. These politicians have blood on their hands, and the death toll continues to rise."

Invoking Nazi Germany with talk of "death panels"

In 2009, 85-year-old U.S. Senator Chuck Grassley of Iowa—who has been in Congress for 39 years—shamefully tried to scare senior citizens and others. He defended his opposition to the Affordable Care Act by saying "we should not have a government program that determines you're going to pull the plug on Grandma." He added "you have every right to fear!" Never mind the fact that private insurance companies already decide whether or not you will get medical treatment, as many procedures require pre-authorization.

Despite polls which show the Affordable Care Act has actually increased in popularity—even amongst Republican voters—most of the leadership of the GOP would prefer to ignore tens of millions of uninsured people by sabotaging a plan that, since its enactment, has made health care coverage available to over 23 million more Americans. These pols are just that craven and corrupt. Incidentally, the ACA has extended improvements even to those who do not buy insurance in the subsidized marketplace. For example, the ACA now prevents insurance companies from denying coverage or charging more for pre-existing conditions. (These include cancer, diabetes and even pregnancy). That benefit alone helps 129 million Americans who otherwise would lose their existing coverage. It is literally a matter of life or death.

(*Opensecrets.org* claims that the Number One industry that has funded Grassley's political campaigns is insurance, that has given him at least $1.5 billion, followed by health professionals, lobbyists, pharmaceuticals/health products, hospitals, health services/HMOs. So "Chuckie the Clown" Grassley—as he is derisively referred to by many Iowans grown weary of his aw-shucks, down home persona that conceals his ruthless pursuit of a rightwing agenda—knows how to appeal to his benefactors, even if it means lying to the voters).

It is hard to fathom that any politician in America who got elected would turn his or her back on the overwhelming crisis that calls for a comprehensive blueprint—such as a Green New Deal—to address the multi-faceted problems the world is confronting today.

Yet there was U.S. Senator Joni Ernst from Iowa—just like most of her fellow Republicans—using the same outrageous and deceitful talking points from the GOP playbook to try to persuade voters not to support action against climate change!

Ernst cited a bogus figure she derived from a Republican so-called "think tank" to assert that "At $93 trillion, the Green New Deal would cost more than the entire recorded spending of the U.S. since the Constitution went into effect in 1789." Her claim was rated "false" by Politifact at the Pulitzer Prize-winning Poynter Institute, as well as the respected website, *Politico*. But it's more than that: it's laughably absurd and any self-respecting politician who had a conscience would be embarrassed to spew such ridiculousness.

Predictably, other conservative outlets—*Forbes*, Fox News, The Heritage Foundation, *Reason*, and even the *Missouri Times*—have all parroted the erroneous $93 million figure.

Normally milquetoast and religious zealot Vice President Mike Pence went on the attack, proclaiming indignantly "The only thing green about the so-called Green New Deal is how much green it's going to cost taxpayers if these people ever pass it into law. You know Margaret Thatcher probably said it best: 'The trouble with socialism is that you eventually run out of other people's money,'" Pence told a cheering crowd recently at the Conservative Political Action Committee (CPAC) Conference. CPAC is the biggest assembly of goons, grifters, and paranoiacs this side of an Infowars taping.

For Pence, an avowed Christian, to equate the Green New Deal with socialism is the sort of deceptive branding Republicans hope to establish in the minds of voters—and they know it is an iniquitous lie.

Other CPAC speakers included "My Pillow" huckster Mike Lindell, who called Donald Trump "the greatest President in history." (Lindell has faced lawsuits over deceptive ads, tax evasion, breach of contract and in 2017 his Better Business Bureau rating was "F").

How could these Green New Deal proposals be controversial?

Let's briefly examine again the main goals of the Green New Deal. Do they really seem so unreasonable? These include:
 • Meeting 100 percent of the power demand in the United States "through clean, renewable, and zero-emission energy sources."

• Eliminating greenhouse gas emissions from U.S. transportation, manufacturing, and agricultural sectors "as much as technologically feasible."

• Upgrading existing buildings and constructing new buildings to achieve "maximum energy efficiency."

• "Working collaboratively with farmers and ranchers in the United States to eliminate pollution and greenhouse gas emissions to achieve "maximum energy efficiency."

The GND advocates transitioning the United States to use 100 percent renewable, zero-emission energy sources, including investment in electric cars and high-speed rail systems, and implementing the "social cost of carbon" that has been part of former President Barack Obama's plans for addressing climate change within ten years. Besides increasing state-sponsored jobs, the Green New Deal is also designed to decrease poverty by aiming much of the improvements in the "frontline and vulnerable communities" which include the poor and disadvantaged among us—and there are many. The GNC also includes calls for universal health care, increased minimum wages, and preventing monopolies.

Although the phrase "Green New Deal" has come to be associated in the public mind with progressives and Democrats, it was actually coined in 2007 by a self-described centrist and "free market guy," *New York Times* columnist Thomas Friedman. Writing in that paper, he called for an end to fossil fuel subsidies, proposed taxing carbon dioxide emissions, and creating lasting incentives for wind and solar energy.

Friedman's ideas made it into the mainstream the following year when Presidential candidate Barack Obama embraced the proposals. In 2009, the United Nations drafted a report calling for a Global Green New Deal to focus government stimulus on renewable energy projects. A month later, Democrats' landmark cap-and-trade bill—designed to set up a market where companies could buy and sell pollution permits and take a conservative first step toward limiting carbon dioxide emissions—passed in the U.S. House of Representatives with the promise of stimulating $150 billion in clean energy investments and creating 1.7 million good-paying jobs.

Friedman's column had focused on policies that compelled the "big players to do the right thing for the wrong reasons." He liked a lot of what Obama and the Democratic-controlled Congress enacted, including $51 billion in "green stimulus" and a $2.3 billion tax credit to clean energy manufacturing—even after Obama stopped using the phrase Green New Deal after the midterm elections.

Unfortunately, by 2010, austerity politics took hold. Republicans eager to block any and all initiatives proposed by Mr. Obama—even ones they had previously supported—relentlessly rejected his legislative initiatives and convinced just enough Democrats to do likewise. GOP Congressional leaders Mitch McConnell and Paul Ryan led their party to oppose everything on the premise that it would add to our national deficit—a concern they quickly and hypocritically abandoned when Republican Donald Trump came into office and exploded the deficit with massive tax cuts for the rich.

Thus the cap-and-trade bill, known as the American Clean Energy and Security Act, died in the Senate in 2010.

In Britain, the Labor Party, acting on a proposal that a team of economists calling themselves the Green New Deal Group drafted, established a government-run green investment bank to bolster renewable energy. But then the conservative Tories swept into office months later and began privatizing the nascent institution. Balanced budgets and deficit hysteria became the dogma of governments across the developed world. Talk of a Green New Deal withered on the vine.

Still, subsidies for wind, solar, and battery technology managed to survive proposed cuts in the tax bill Congress passed in 2017 after Trump took office, because Republicans in states that have come to rely on those burgeoning industries saved them. For Friedman, that is proof that lasting climate policies are ones that make private renewable energy companies powerful enough to sway politics.

"The more the market does on its own, the more sustainable it is," Thomas Friedman says. He is skeptical of government intervention to wean the nation off fossil fuels, and believes the goals could be achieved merely by messaging that focuses on the patriotic, nation-building aspects of greening the economy.

Friedman insists "We are the true patriots on this. We're talking about American economic power, American moral power, American geopolitical power. Green is geostrategic, geoeconomic, patriotic, capitalistic."

But in this era when messages are overridden daily by a corrupt, narcissistic, shameless, petty, vindictive, bullying ignoramus and lying loudmouth, Donald Trump, it's hard to penetrate the noise and get sensible ideas presented to the public arena. More than rhetoric is needed; legislative action must be pursued to implement important reforms to deal with environmental issues.

After Obama stopped using the term, there was "radio silence" about a Green New Deal until it came to life again during the Democratic campaign of Senator Bernie Sanders for President in 2016.

In a historic upset, Democrats regained control of the House of Representatives in the 2018 mid-term elections. Led by Nancy Pelosi, Dems gained 41 seats—their largest gain of House seats since the post-Watergate 1974 elections, when they picked up 49 seats. The Democrats also won the popular vote by a margin of 8.6 percent, the largest margin on record for a party that previously held a minority in the House. Turnout was the highest for a midterm election in more than a century, with over half the electorate casting ballots.

Many of the Democratic newcomers to Congress have unabashedly embraced progressive ideas ranging from supporting health care to the Green New Deal.

No one exemplifies this better than the 29-year-old rising political star, Alexandria Ocasio-Cortez, who shocked everyone in the 2018 mid-term elections by ousting Democratic incumbent Representative Joe Crowley in a working-class Bronx and Queens district in New York City, in part because she outlined a vivid version of the GND. Her upset victory is all the more amazing because Crowley had been the fourth-most powerful Democrat in Congress and possible House speaker who had deep ties to corporate interests and funding.

Rep. Alexandria Ocasio-Cortez is a leading advocate for the Green New Deal

By December 2018, a survey by Yale and George Mason universities found 81 percent of American voters supported the goals of a Green New Deal, including 64 percent of Republicans and 57 percent of conservative Republicans. Yet the pollsters warned that support would likely wane as the policy became more closely associated with members of one party.

On January 10, 2019, a letter signed by 626 organizations in support of a Green New Deal was sent to all members of Congress. It called for measures such as "an expansion of the Clean Air Act; a ban on crude oil exports; an end to fossil fuel subsidies and fossil fuel leasing; and a phase-out of all gasoline-powered vehicles by 2040."

Desperation on the right

Panicking about how popular the Green New Deal has become virtually overnight, the far-right has launched their paranoiac attacks which have already reached a fever pitch and are likely to escalate as the Presidential election year of 2020 arrives.

Predictably, Fox News, Breitbart News and such rightwing media as the *Las Vegas Review-Journal* have spread inaccurate claims about the Green New Deal in a frantic attempt to stoke fears that it would destroy the American way of life. The Heartland Institute, a far-right "think tank" (an oxymoron if there ever was one), is known for its climate denial. Justin Haskins, an executive editor and research fellow for Heartland and author of the book *Socialism Is Evil*, is the group's main spokesman for its claim that the Green New Deal has a hidden socialist agenda.

Haskins asserts that the real goal of the GND is to impose a series of radical, socialistic programs—policies that would dramatically increase the size and power of the federal government, cause immense harm to the U.S. economy, and run up the national debt by trillions of dollars. He warns "If we don't stop it, it will destroy our economy for a whole generation of Americans."

In a January 3, 2019 *Washington Examiner* op-ed, Haskins claimed, "This is one of the most dangerous and extreme proposals offered in modern U.S. history. It's the sort of thing you'd see in the Soviet Union, not the United States."

Justin Haskins (right) appears on FOX's Tucker Carlson show to claim the Green New Deal is extreme socialism — a falsehood.

Heartland's attacks on the Green New Deal are par for the course, because the organization has a long history of promoting climate denialism. Heartland exists to push a hardcore platform of deregulation, austerity and anti-labor policies. It has received funding from groups backed by the oil billionaires, the Kansas-based Koch brothers, whose father co-founded the radical John Birch Society years ago that was so extreme in its views—and racist—that it was disavowed by conservative William F. Buckley, Jr. Heartland has also received funding from fossil fuel companies, but none of this was disclosed in any of Haskins' recent Fox appearances or op-eds attacking the Green New Deal.

Ramping up the hysteria, Breitbart claims the GND traces back to Karl Marx.

Other far-right voices have joined the chorus. Tom Elliott argued that the plan amounts to a "radical grant of power to Washington over Americans' lives, homes, businesses, travel, banking, and more." Jarrett Stepman wrote that the plan "would upend our way of life and destroy the liberty and prosperity" that we enjoy. Jim Geraghty wrote that "enacting these changes would probably require a dictatorship or other authoritarian regime." Brian Mark Weber called the plan "eco-fascist." During an appearance on *Fox & Friends*, Fox News contributor Tammy Bruce compared the GND to "economic enslavement."

Fox feels so threatened by Rep. Alexandria Ocasio-Cortez, they mentioned her derogatorily on their news and business channels 3,181 times in a recent 41-day period. Obviously, the propaganda network is trying to brainwash their audience into viewing the Green New Deal with alarm.

Fox's Tucker Carlson called Ocasio-Cortez an "idiot wind bag," a "pompous little twit," a "fake revolutionary," "self-involved and dumb," a "moron and nasty and more self-righteous than any televangelist." Of course, Carlson routinely makes offensive and misogynistic comments, (eg., women are "extremely primitive, they're basic, they're not that hard to understand.") He even defended a Mormon minister currently serving a life sentence for child sexual assault.

Responding to the harsh caricatures of her by rightwing critics, Ocasio-Cortez said she expected the criticism of the Green New Deal plan but added, "I didn't expect them to make total fools of themselves."

The promotional blurb for a climate change denial book by Martin Capages Jr. is more reflective of the sort of mentality evinced by members of the Flat Earth Society than it is anyone who has even a modicum of insight into what the scientific evidence actually is.

Yet Capages' book promo ludicrously states that "the entire concept [of the Green New Deal] is based on the false premises that carbon dioxide (CO2) is a pollutant, CO2 emissions are bad, that the increasing rate of CO2 emissions from burning fossil fuels is causing the global temperature to rise which will lead to adverse climate change, and that reducing fossil fuel use is an immediate necessity to protect the natural environment from most human actions. None of the premises are true but an entire political party has been co-opted by the 'old' glamour of young, naïve want-to-be Marxists who have seized on this invalid concept, The New Green Deal, to disguise a push for complete State control in all matters. The Party establishment has caved and now Democratic Socialism is a primary plank in the Party's progressive political platform."

Such misleading rhetoric is reflective of how scared rightwing groups and their backers in the fossil fuel industry are about the Green New Deal, and justifiably so. Because by March 2019, even as the partisan polarization took effect and Donald Trump joined the attack on the GND, the popularity remained. Fifty-two percent of adults supported the Green New Deal, compared to 42 percent who disagreed with it, according to an NBC News poll released just before this book went to press.

"It's more popular than the Republican tax cuts," noted Heather McGhee, a senior fellow at the progressive think tank Demos. "We have this idea that left ideas are moving and shaking inside the political atmosphere right now. That's true. But they're left in terms of Washington. But in terms of the American people, they're centrist ideas."

In late March 2019, Rep. Alexandria Ocasio-Cortez (D-N.Y.) entered the auditorium at the Albert Einstein College of Medicine to a standing ovation as she arrived to tape an hour-long TV special on the Green New Deal, her signature policy proposal, with MSNBC's Chris Hayes.

She said she understood "why people say [the GND represents] 'the Tea Party of the left.'" It has surprising momentum across the political spectrum and is growing exponentially. But, Rep. Ocasio-Cortez explained, "This is not the Tea Party of the left, this is a return to American representative democracy. The big difference? The Koch brothers funded the Tea Party. Everyday people funded my campaign."

Besides the environmental-friendly aspects of the GND, some proponents want to combine this with the proposal of guaranteeing a job with a family-sustaining wage, high quality health care, adequate housing and retirement security to all people of the United States.

Rep. Ocasio-Cortez also addressed the political economic components of the plan by saying "We cannot allow for the fossil fuel jobs to be better, more dignified and higher-waged with stronger labor support behind [them] than new energy jobs. What I'm tired of is us worrying more about the future of fossil fuels than worrying about the future of fossil fuel workers."

Ocasio-Cortez Tweeted that sewing "discord and scarcity is an effective tool for billionaires and the powerful, because those are the most effective ways to turn working people against one another. Solidarity, fighting for each other, and addressing injustice is our path to economic dignity."

Indeed, the whole subject of a changing climate and how to respond to it, tied in with the despoilation of our natural resources—wilderness areas, forests, oceans—cannot be addressed without also noting the political context, including the rising white supremacy movement and fortressed borders, or what Naomi Klein calls a form of "climate barbarism."

Corporate shareholders respond

Not surprisingly, the economic safety net and social justice components of the Green New Deal have antagonized the extremists of the right, who historically have viewed any proposal that actually helps the poor, disadvantaged or even middle income Americans as encroaching on their self-appointed status as the Gilded Class. Most of the super rich seem to be missing an empathy gene, and their unbridled greed has propelled them to use their considerable financial resources and influence over politicians and far-right media to effectuate public policies that harm the majority of Americans so the elite can enjoy unchallenged lives of privilege and domination. The histrionics from the rightwingers are an indication that even the vague Green New Deal has touched a raw nerve, as can be seen in the preposterous exaggerations about the proposal that have been made. The radical American Action Forum, for example, claims the potential cost of the GND at $600,000 per household.

Mitch McConnell staged a stunt by scheduling a vote in the Senate on the Green New Deal before the proposal was fleshed out, and it predictably failed. *Forbes* announced the GND dead.

Yet according to *Bloomberg Businessweek*, Wall Street is amenable to investing significant financial resources in a Green New Deal.

Axios reports that investors in big corporations have been presenting "non-binding but symbolically important resolutions to companies on a range of topics," with "environmental issues, particularly climate change, [topping] the list for activist investors, account[ing] for 21 percent or roughly 400 proposals filed." Why it matters is, as Axios explains, "the investment community is becoming an alternative battleground between big companies and climate change as government policy on the matter retreats under Trump. While the resolutions are nonbinding, companies usually comply with the requests if [the proposed resolutions] get more than 50 percent support."

The New York public pension fund and Church of England's endowment filed a resolution for consideration at Exxon's annual meeting calling for the company to disclose targets that would drastically reduce its greenhouse gas emissions.

Yet ExxonMobil has attempted to throw out an investor proposal calling on the oil giant to set targets to reduce its greenhouse gas emissions.

Investing in green initiatives makes sense. But there is a significant gap between the capital that must be applied to tackle climate change and the amount that is being deployed today. So in a laudable move, Bank of America announced an additional $300 billion dollar environmental business commitment to increase the flow of capital to low-carbon business activities and advance the United Nations Sustainable Development Goals.

Existing programs training workers in green skills include a program called Roots of Success, founded in 2008 to bring low-income people into living wage professions. Funding for Roots of Success came from the $90 billion in green initiatives incorporated in the American Recovery and Reinvestment Act [ARRA] of 2009.

About 12 percent of ARRA funding went to green investment, and some of these initiatives were successful. A January 2019 article in *Politico* stated that "U.S. wind capacity has more than tripled since 2008, while solar capacity is up more than sixfold. LEDs were one percent of the lighting market in 2008; now they're more than half the market. There were almost no plug-in electric vehicles in 2008; now there are more than one million on U.S. roads."

Although ARRA's green stimulus projects are of interest for developing proposals for a Green New Deal, its mixed results included both "boosting innovative firms" such as Tesla, and the $535 million failure of the Solyndra solar company." These initial efforts at green stimulus are described as a "cautionary tale."

"The climate is a common good, belonging to all and meant for all."
—Pope Francis

Religious leaders speak out

Sustainable development cannot be achieved without the voices of those effected by the exploitation of the earth's resources, especially the poor, migrants, indigenous people and young men and women, Pope Francis told participants at a Vatican conference on sustainable development in March 2019.

Without a change of attitude that focuses on the well-being of the planet and its inhabitants, efforts to achieve the U.N.'s sustainable development goals will not be "sufficient for a fair and sustainable world order," the Pontiff said.

"Religions have a key role to play in this," the Pope asserted. "For a correct shift toward a sustainable future, we must recognize our errors, sins, faults and failures, which leads to a heartfelt repentance and desire to change; in this way, we will be reconciled with others, with creation and with the Creator."

The three-day international conference, titled "Religions and the Sustainable Development Goals: Listening to the cry of the earth and of the poor," looked at how religions can help the world reach the goals by 2030. The meeting brought together religious leaders from all major faith backgrounds as well as advocates and experts in the fields of development, the environment and health care.

Development, he said, is "a complex concept" that has been "almost entirely limited to economic growth," thus leading the world down a "dangerous path where progress is assessed only in terms of material growth." Such a path has caused many to "irrationally exploit the environment and our fellow human beings," he added.

Pope Francis urged that we "commit ourselves to promoting and implementing the development goals that are supported by our deepest religious and ethical values. Human development is not only an economic issue or one that concerns experts alone; it is ultimately a vocation, a call that requires a free and responsible answer."

Francis also expressed his hope that the conference would lead to concrete solutions that respond to "the cry of the earth and the cry of the poor" while promoting serious commitments "that develop alongside our sister earth and never against her.

"If we are truly concerned about developing an ecology capable of repairing the damage we have done," the Pope said, "no branch of science or form of wisdom should be overlooked, and this includes religions and the languages particular to them."

Although the non-specific nature of current GND proposals has become a concern for some Greens, one writer from the Columbia University Earth Institute views the lack of specificity as a strength, noting that: "FDR's New Deal was a series of improvisations in response to specific problems that were stalling economic development. There was no master plan, many ideas failed, and some were ended after a period of experimentation. But some, like Social Security and the Security and Exchange Commission's regulation of the stock market, became permanent American institutions."

Actually, as *Time* magazine reflects, today's Green New Deal is not dissimilar to Roosevelt's original New Deal: "The modern calls for focusing on upgrading infrastructure and for 'smart' power grids echo the work of the original New Deal's Rural Electrification Administration, which brought electricity to rural areas. The call for 'reducing the risks posed by flooding' recalls the Tennessee Valley Authority (TVA) supervising the construction of dams for flood control, and the call for construction that safeguards people from extreme weather events recalls the planting of a 'shelter belt' of trees to break up the powerful winds that contributed to the Dust Bowl. And the New Deal's Civilian Conservation Corps (CCC) employed about three million men in what might today be called green jobs, including carving park trails and fighting wildfires. When the CCC disbanded in 1942, *Time* said it had 'one of the best records of all New Deal hopefuls' and that '[the] most rabid opponents of New Deal spending admitted that CCC was worthwhile.'"

The magazine notes that "These programs made a big difference for the land, as well as for the people they employed: A 1946 report found that New Deal projects built or upgraded 650,000 miles of roads, more than 125,000 buildings, more than 16,000 of "water mains and distribution lines," and more than 24,000 miles of sewerage facilities. Also thanks to the New Deal, nearly 220 million trees were planted by the 1940s and about nine out of ten farms had electricity by 1945, according to the Roosevelt Institute."

Those numbers may be seen by advocates as a boost for the possibilities of a Green New Deal, says David Woolner, Senior Fellow and Hyde Park Resident Historian at The Roosevelt Institute and co-editor of *FDR and the Environment*. After all, they're proof that environmental change on a grand scale is possible under the circumstances.

"We tend to think these green-jobs programs are beyond the reach until our economy is up and running at a proper pace—too expensive, can't afford it," Woolner says. And yet the New Deal was implemented at "the worst period of economic crisis in America's history."

We need citizen action to make changes

Now, as in the 1930s when the first New Deal was adopted, we can't rely on corporate America to make the changes needed.

Rep. Alexandria Ocasio-Cortez points out, "ExxonMobil knew that climate change was real and man-made as far back as 1970. The entire United States government knew that climate change was real and human-caused in 1989, the year I was born. So, the initial response was, 'Let the market handle it, they will do it.' Forty years, and free-market solutions have not changed our position. So, this does not mean that we change our entire structure of government. What it means is that we need to do something—*something*—and that is what this solution is about."

She observes, "To get us out of this situation, to revamp our economy, to create dignified jobs for working Americans, to guarantee health care and elevate our educational opportunities and attainment, we will have to mobilize our entire economy around saving ourselves and taking care of this planet."

Rep. Sean Duffy (R-Wis.) accused the Democrats of touting policies that he claimed would solely help "rich liberals," but alienate poorer Americans.

"I think it's rich that we talk about how we care about the poor, but all the while we'll sign on to bills that will dramatically raise the cost for a family to get into a home," Duffy said while speaking about a measure related to homelessness. He later called the Green New Deal "absolutely outrageous."

That prompted a fierce retort from Ocasio-Cortez. "This is not an elitist issue, this is a quality of life issue," the Congresswoman insisted. "You want to tell people that their concern and their desire for clean air and clean water is elitist? Tell that to the kids in the south Bronx which are suffering from the highest rates of childhood asthma in the country. Tell that to the families in Flint."

Democratic political consultant David Axelrod says climate change didn't resonate with voters at the start of the Obama administration, but "Now we're a decade down the road, and the road is surrounded by floods and fires in a way that is becoming more and more visible."

Encouragingly, younger members of Congress may be able to find common ground and help build a consensus for action.

A Republican Rep. from Florida, Matt Gaetz—known for his brash, outspoken nature—surprisingly told the *New York Post* that he and Rep. Ocasio-Cortez "actually...had a very productive conversation about our areas of agreement."

The New York rep had attacked Republicans for supporting subsidies to the oil and gas industry, saying "I would also like to highlight that it is not responsible to complain about anything that we dislike as quote-unquote socialism. Particularly when many of our colleagues across the aisle are more than happy to support millions and potentially billions of dollars in government subsidies and carve-outs from the oil and gas and fossil fuels industry," she said.

Gaetz said he would like to see them gone as well. "If [she] wants to file a bill to end subsidies to the fossil fuel industry, I'll be the first co-sponsor," Gaetz said. "That would be a good thing for the climate change agenda and I would support her on that."

He said the two also see eye to eye on trade and intellectual property.

"We both agree that U.S. innovators shouldn't be robbed blind by China and that's what happens in solar and it's happening right now in electric cars," Gaetz said.

"We both agree that our current electric grid sucks and needs major improvement," he added. We "share the same objective... She is more of a public investment, I prefer a blend with private investment" to update the grid.

Gaetz, 36, and Ocasio-Cortez, 29, are similar in that they've used their big personalities and social media presence to get attention.

Recently, Gaetz introduced his "Green Real Deal," which is similar to the "Green New Deal" in that it outlines a strategy to combat climate change.

Matt Gaetz says he prefers "a more modest role for government and...a more robust embrace of American innovation."

The biggest problem, Gaetz said, is on Capitol Hill, there's "no acceptance of a strategy."

Urgent problems facing us

Writing in *Vox*, David Roberts outlines the scope of what he sees as a climate "emergency": "The conservative movement and the Republican Party have descended into unrestrained tribalism, rallying around what is effectively a crime boss who it now appears was elected with the help of a hostile foreign power. The media has calved in two, with an entire shadow right-wing media capturing the near-exclusive attention of movement conservatives, descending into increasingly baroque and lurid fantasies.

"The President is now openly admitting to scheduling a 'national emergency' because he wanted money for his wall, itself a lurid xenophobic fantasy. Meanwhile he is doing everything in his power to delay or shut down multiple federal investigations into his possible crimes. At every stage of his descent into paranoid lawlessness he has had the support of Republicans in Congress (because he lowered taxes on rich people) and the near-unanimous backing of Republican voters (because he owns the libs).

"Basic norms of political conduct are crumbling on a daily basis. The country's core institutions are under intense stress. It plays out on television and social media like an exhausting spectacle, always turned to 11.

"And it all takes place in the context of Americans' shrinking faith in their political system, which is ever-more-nakedly funneling wealth and power to the already wealthy and powerful (while protecting them from accountability) and heaping more risk and instability onto the most vulnerable.

"The reactionary (largely older white male) backlash and the rising appeal of democratic socialism among the young are both, in their own ways, responses to a money-soaked, unresponsive political system." That pretty well sums it up.

3.
From Early Settlers to Modern Times

FROM THE ARRIVAL OF CHRISTOPHER COLUMBUS STARTING IN 1492, and through the Age of Exploration when Europeans began to come to America in large numbers and overtake the indigenous Native American peoples, the area that later became the United States has been a land plundered for its natural resources and otherwise abused. As settlements were established here principally by Spain, England and France, rivalry among these European powers created a series of wars on the North American landmass that would have great impact on the development of the colonies. Territory often changed hands multiple times. During Pontiac's Rebellion between the years 1763-1766, a confederation of Great Lakes-area tribes fought a somewhat successful campaign to defend their rights over their lands west of the Appalachian Mountains. But ultimately, with the formation of the United States, the newly minted nation expanded rapidly to the west and supplanted the indigenous people who had sovereignty over the land for centuries.

This was a great paradox because the early white settlers had come here to pursue liberty and religious freedom, yet somehow rationalized their subjugation and displacement of the native population, depriving them of those same rights. Like the African slaves some of the white Europeans brought to America, they didn't feel "the red man" was on equal terms with caucasians.

George Washington is best known as the first President of the United States, after having been Commander-in-Chief of the Continental Army during the Revolutionary War. But one of his proudest personal achievements is how he treated the environment and improved the productivity of his land at his estate, Mt. Vernon.

Those closest to him believed Washington was happiest when he was working his land and conducting agricultural experiments. He was primarily a tobacco farmer, but eventually diversified into growing wheat, corn, carrots, cabbage, and a variety of other crops. He also used the results to determine what would grow best in his type of soil. He experimented with a seven-year crop rotation and in numerous diaries, essays and speeches, Washington encouraged Americans to enrich their land instead of wearing it out. So he could be regarded as one of the first conservationists in the country.

Thomas Jefferson, third President of the United States, was probably our most accomplished man in public life as well as the most versatile. He was considered an expert in many areas: architecture, civil engineering, geography, mathematics, ethnology, anthropology, mechanics, and the sciences. During his lifetime, he was an infallible oracle to half the population of the country and a dangerous demagogue to the other half, but was universally recognized as a man of political and literary attainments. Jefferson was so familiar with every subject discussed by ordinary people and talked so fluently and with such confidence, that he was considered a monument of learning by his fellow countrymen. An inveterate note-taker, he jotted down in journals things he had learned from conversations with others. About the only attribute Jefferson failed to possess was a sense of humor. He was utterly immune to witticisms or humorous stories.

Yet, like Washington, Thomas Jefferson kept detailed notes on his stewardship of the land—his estate called Monticello. Believing that "the greatest service which can be rendered any country is to add a useful plant to its culture," he imported plants and trees from abroad and tried growing olives, oranges, almonds and French grapes at Monticello. So eager was he to improve the agriculture of his country that when Minister of France he broke the law by personally smuggling rice from Italy. This small store of seeds became the basis for the South Carolina rice industry.

During his Presidency, the U.S. acquired the massive Louisiana Territory in 1803, and the next year, Thomas Jefferson dispatched Meriwether Lewis and William Clark to explore lands west of the Mississippi, to the Pacific Ocean. Their 8,000 mile journey lasted over two years, during which their party of 33 people encountered harsh weather, rugged terrain, treacherous waters, and Native Americans who were at times friendly but otherwise hostile. They brought with them beads, face paint, knives, tobacco, ivory combs, brightly colored cloth, ribbons, sewing notions and mirrors with which to gift the Native Americans they encountered along the way. In meeting new tribes, Lewis and Clark bartered goods and presented the tribe's leader with

The Jefferson Indian Peace Medal

a Jefferson Indian Peace Medal—a coin engraved with the image of the President on one side, and the image of the clasped hands of an Indian and a U.S. soldier, beneath a tomahawk and a peace pipe on the other, with the inscription "Peace and Friendship."

However, the explorers told the Indians that America *owned* their land and offered military protection in exchange for peace. Some Indians had met "white men" before and were friendly and open to trade. Others were wary of Lewis and Clark and their intentions and were openly hostile—though seldom violent.

The nation rejoiced when Lewis and Clark and their men— rumored to be dead or lost—safely returned to the U.S. Capitol. The explorers provided Thomas Jefferson with new geographic, ecological and social information about the previously uncharted areas of North America. They identified at least 120 animal species and 200 botanical samples.

In 1830, Congress passed the so-called Indian Removal Act, which forced Native Americans to move from the Southeast to west of the Mississippi River. In 1838, the Cherokee Nation was forced to march from the east coast to Oklahoma, and many thousands died along the way. Commencing in 1841, Americans began to travel west in wagon trails on the Oregon Trail. About 300,000 would follow this trail over the next 20 years.

Between 1846 and '48, a war was fought over the rights to Texas (the "Mexican-American War"), after which the United States paid Mexico $15 million for land that would later become California, Texas, Arizona, Nevada, Utah, and parts of several other states. In the midst of this, England signed the Oregon Treaty handing over the Oregon Territory to the United States.

By 1849, the "Gold Rush" had begun, when many Americans hurried to California hoping to strike it rich. The first transcontinental telegraph was thereafter established, and then the federal government decided to fund the building of a Transcontinental Railroad from Missouri to California—1,776 miles in length.

The Civil War's environmental impact was devastating

The Civil War, also known as "The War Between the States," (April 12, 1861 to April 9, 1865), was fought between the United States and the Confederate States—the latter comprising a collection of 11 southern states that left the Union in 1860 and 1861 and formed their own country in order to perpetuate slavery. We tend to think of the war's devastating effects only in terms of lives lost (620,000). But it was also an environmental catastrophe of the first magnitude, with effects that prevailed long after the guns fell silent.

Both the Union and Confederate armies pursued a "scorched earth" policy, laying waste to farms, fields, crops and timber. Encampments fouled the water and spread disease. Insects thrived in the camps because the decimated forests had been the habitats of the birds and bats and other predators that would have kept the pest populations down. More than 1.3 million soldiers in the Union alone were affected by mosquito-borne illnesses like malaria and yellow fever.

Although Henry David Thoreau died in 1862, before the war was over, he borrowed from some of its imagery to bemoan a "war on the wilderness" that he saw all around him. His final manuscripts suggest that he was working on a book about the power of seeds to bring rebirth—not dissimilar to what Abraham Lincoln would say in the Gettysburg Address.

Lincoln's minister to Italy, George Perkins Marsh, wrote a book in 1864 called *Man and Nature*, in which he

Two-thirds of the deaths during the Civil War were due to disease; dysentery and typhoid fever were the biggest killers of the men, not battle.

In the midst of the Civil War, the visionary President Abraham Lincoln was thinking ahead about preserving our natural wonders by protecting the Yosemite area.

condemned the obliviousness of Americans to the ways in which they were exploiting and destroying nature without thought given to tomorrow.

On June 30, 1864, Abraham Lincoln signed a bill to set aside land known as Yosemite and giving it to the state of California with the stipulation it "shall be held for public use, resort, and recreation" and shall, like the rights enshrined by the Declaration, be "inalienable for all time." As the nation was roiled by the Civil War, this was a unique concept for the time, and attracted little attention. However, it set a precedent whereby scenic land could be preserved for the enjoyment of future generations. Lincoln also founded the Department of Agriculture, which later included various iterations of the U.S. Forest Service.

President Lincoln

Southerners also expressed their reverence for nature. Gen. Robert E. Lee wrote his wife, after traveling through what is now

West Virginia, "I enjoyed the mountains, as I rode along. The views are magnificent—the valleys so beautiful, the scenery so peaceful. What a glorious world Almighty God has given us. How thankless and ungrateful we are, and how we labour to mar his gifts."

Another development that occurred during the Civil War years has been recounted by Ted Widmer, writing in *The New York Times*. He noted that "As the world's whale population began to decline in the 1850s, a new oily substance was becoming essential. Petroleum was first discovered in large quantities in northwestern Pennsylvania in 1859, on the eve of the war. As the Union mobilized for the war effort, it provided enormous stimulus to the new commodity, whose uses were not fully understood yet, but included lighting and lubrication. Coal production also rose quickly during the war. The sudden surge in fossil fuels altered the American economy permanently. Every mineral that had an industrial use was extracted and put to use, in significantly larger numbers than before the war."

Interestingly, the first scientific attempt to explain heat-trapping gases in the earth's atmosphere and the greenhouse effect was made in 1859 by an Irish scientist, John Tyndall. But it would be many decades before his research was taken seriously.

The passenger pigeon is now extinct.

Some species were wiped out by the Civil War. One example is the passenger pigeon. The bird was once so abundant that flocks of them could darken the sky, flying at speeds of up to 62 miles per hour! The male was mainly gray on the upperparts, lighter on the underparts, with iridescent bronze feathers on the neck, and black spots on the wings. Deforestation destroyed its habitat, and it was forced to relocate westward. But a culture of game-shooting developed in the years right after the war, accelerated not only by widespread gun ownership, but by a supply-and-demand infrastructure developed during the war, along the rails. When Manhattan diners wanted to eat pigeon, there were always hunters in the upper Midwest willing to shoot at boundless birds—until suddenly the birds were gone. They declined from billions to dozens between the 1870s and the 1890s. One hunt alone, in 1871, killed 1.5 million birds. Another, three years later, killed 25,000 pigeons a day for five to six weeks. The last known passenger pigeon, Martha, died at the Cincinnati Zoo on Sept. 1, 1914.

The United States government has identified 384 battles that had a significant impact on the larger war. Many of these battlefields have been developed—turned into shopping malls, pizza parlors, housing developments, etc.—and many more are threatened by development.

However, a lot still remain. As Ted Widmer reflects, "Paradoxically, there are few places in the United States today where it is easier to savor nature than a Civil War battlefield. Thanks to generations of activism in the North and South, an extensive network of fields and cemeteries has been protected by state and federal legislation, generally safe from development. These beautiful oases of tranquility have become precisely the opposite of what they were, of course, during the heat of battle. They promote a reverence for the land as well as our history, and in their way, have become sacred shrines to conservation.

"Perhaps we can do more to teach the war in the same way that we walk the battlefields, conscious of the environment, using all of our senses to hear the sounds, see the sights and feel the great relevance of nature to the Civil War. Perhaps we can do even better than that, and summon a new resolve before the environmental challenges that lie ahead. As Lincoln noted, government of the people did not perish from the earth. Let's hope that the earth does not perish from the people."

The Transcontinental Railroad was a mixed blessing

As the *Smithsonian* magazine reports, "Not long after President Abraham Lincoln signed the Pacific Railway Act of 1862, railroad financier George Francis Train proclaimed, 'The great Pacific Railway is commenced. Immigration will soon pour into these valleys. Ten millions of emigrants [sic] will settle in this golden land in 20 years. This is the grandest enterprise under God!' Yet while Train may have envisioned all the glory and the possibilities of linking the East and the West coasts by 'a strong band of iron,' he could not imagine the full and tragic impact of the Transcontinental Railroad, nor the speed at

The Transcontinental Railroad was both a blessing and a curse.

which it changed the shape of the American West. For in its wake, the lives of countless Native Americans were destroyed, and tens of millions of buffalo, which had roamed freely upon the Great Plains since the last ice age 10,000 years ago, were nearly driven to extinction in a massive slaughter made possible by the railroad."

Chief Red Shirt wearing one of the U.S. government peace medals.

Construction of the railroad had an unfortunate side effect in that it had dire consequences for the native tribes of the Great Plains, forever altering the landscape and causing the disappearance of once-reliable wild game. The railroad was probably the single biggest contributor to the loss of the bison, which was particularly traumatic to the Plains tribes who depended on it for everything from meat for food to skins and fur for clothing, and more.

In response, Native Americans sabotaged the railroad and attacked white settlements supported by the line, in an attempt to reclaim the way of life that was being brutally taken from them. If they were not taking aim at the railroad tracks and machinery, they would attack the workers and abscond with their livestock. Ultimately the tribes of the Plains were unsuccessful in preventing the loss of their territory and hunting resources. Their struggle serves as a poignant example of how the Transcontinental Railroad could simultaneously destroy one way of life as it ushered in another.

After the Civil War, Gen. William Tecumseh Sherman was appointed to oversee the westward expansion and protect the railroads. In 1867, he wrote to Gen. Ulysses S. Grant, "we are not going to let thieving, ragged Indians check and stop the progress" of the railroads. The last rails were laid in 1869.

The Indian Wars saw U.S. troops launch merciless attacks on Native Americans, slaughtering adults, children and animals. By 1878, Indians had been consigned to reservations with no compensation beyond the promise of religious instruction and basic supplies of food and clothing—promises which were never fulfilled.

Smithsonian magazine explains, "Massive hunting parties began

A pile of bison skulls attests to the wholesale slaughter of this animal to the point of near extinction.

to arrive in the West by train, with thousands of men packing .50 caliber rifles, and leaving a trail of buffalo carnage in their wake. Unlike the Native Americans or Buffalo Bill, who killed for food, clothing and shelter, the hunters from the East killed mostly for sport. Native Americans looked on with horror as landscapes and prairies were littered with rotting buffalo carcasses. The railroads began to advertise excursions for 'hunting by rail,' where trains encountered massive herds alongside or crossing the tracks. Hundreds of men aboard the trains climbed to the roofs and took aim, or fired from their windows, leaving countless 1,500-pound animals where they died."

By the end of the 19th century, only 300 buffalo were left in the wild. Congress finally took action, outlawing the killing of any birds or animals in what later became Yellowstone National Park, where the only surviving buffalo herd could be protected.

Meanwhile, the U.S. government offered farmers free land to live on for five years, with the stipulation they make improvements. These white homesteaders decimated land that had never been previously owned. The Homestead Act of 1862, which provided settlers with 160 acres of public land, was followed by the Kinkaid Act of 1904 and the Enlarged Homestead Act of 1909. These acts led to a massive influx of new and inexperienced farmers across the Great Plains.

Many of these settlers lived by the superstition "rain follows the plow." Immigrants, land speculators, politicians and even some scientists believed that homesteading and agriculture would permanently affect the climate of the semi-arid Great Plains region, making it more conducive to farming. Those beliefs turned out to be disastrously wrong several decades hence.

This false belief was linked to Manifest Destiny—an attitude that Americans had a sacred duty to expand westward. A series of wet years during the period promoted a further misunderstanding of the region's true climate patterns and led to the intensive cultivation of increasingly marginal lands that couldn't be reached by irrigation.

This eventually culminated in the catastrophic "dust bowl" years in the next century...

The first National Park is designated

As more Americans were becoming aware of the rich natural resources the new nation was acquiring, it was President Ulysses S. Grant who, in 1872, designated Yellowstone as the country's first official National Park. To permanently close to settlement an expanse of the public domain the size of Yellowstone rather dramatically departed

These photos by William Henry Jackson in 1871 show Valley of the Yellowstone (top) and the first view of the Old Faithful geyser eruption which fascinated Americans.

from the established policy of transferring public lands to private ownership. But the wonders of Yellowstone had caught the imagination of Congress and Americans in general.

Previously, in the popular imagination, the area was a hellish, wild place that was fraught with dangers from erupting geysers. But in 1871, painter Thomas Moran and photographer William Henry Jackson had joined the first U.S. government survey of the region. For two weeks, Jackson took pictures of the area, and Moran filled his sketchbook—then painted watercolors—of the landscape's most impressive sights. "The photographs were proof that what the artist was showing really existed," says Eleanor Harvey, senior curator at the Smithsonian American

Art Museum in Washington, D.C..

Thus the United States Congress established Yellowstone National Park in 1872. On March 1, 1872, President Ulysses S. Grant signed the Yellowstone National Park Protection Act into law. The world's first national park was born and was then marketed as a wonderland.

Benjamin Harrison preserved an ancient indigenous community

Prior to becoming President for one term in 1889, Benjamin Harrison had already championed the idea of protecting our natural wilderness when he served in the U.S. Senate. He was the first to introduce a bill to create Grand Canyon National Park, variations of which he repeatedly tried to pass thereafter. Once in the White House, he issued an executive order to create the Afognak Island Forest and Fish Culture Reserve in Alaska, today part of Katmai National Park and Preserve, to be "protected and preserved unimpaired." He did so largely to safeguard salmon. Thus Harrison can be credited with creating a true national wildlife refuge.

That same year, Harrison ordered the creation of the Casa Grande Ruin Reservation in Arizona, "the first prehistoric and cultural site to be established in the United States." It is an ancient Sonoran Desert People's farming community. This was a spectacular move on President Harrison's part, which established a precedent for protecting the first prehistoric and cultural reserve in the United States. His set-aside included "the Great House," thought to date back to 1350 B.C. One of the largest prehistoric structures ever built in North America, its purpose remains a mystery.

The Great House in the Sonoran Desert.

Archeologists have discovered evidence that the ancient Sonoran Desert people who built the Casa

Grande also developed wide-scale irrigation farming and extensive trade connections which lasted over a thousand years until about 1450 B.C. There were earthen buildings, red on buff pottery, and extensive canals. Since the ancient Sonoran Desert people who built it left no written language behind, written historic accounts of the Casa Grande begin with the journal entries of Padre Eusebio Francisco Kino when he visited the ruins in 1694.

From the 1860s to 1890s, more people began to visit the ruins with the arrival of a railroad line 20 miles to the west and a connecting stagecoach route that ran right by the Casa Grande. The resulting damage from souvenir hunting, graffiti and outright vandalism raised serious concerns about the preservation of the Casa Grande, so President Harrison took action.

Grover Cleveland built on Benjamin Harrison's eco legacy

Though Grover Cleveland defeated Benjamin Harrison in winning the second of his non-consecutive Presidential terms, he nevertheless continued many of his predecessor's conservation policies.

Using Harrison's Forest Reserve Act, President Cleveland claimed millions of wooded acres to set aside for protection. The most famous of these were 13 new or expanded reserves, enraging timber tycoons, who dubbed them "Midnight Reserves" (conservationists called them the "Washington's Birthday Reserves"). Earlier, in reaction to the lackluster prosecution of poachers in Yellowstone and a public outcry by the Theodore Roosevelt-founded Boone and Crockett Club, in 1894, Cleveland signed into law the Act to Protect the Birds and Animals in Yellowstone National Park. Long after he left the White House, Cleveland boasted of having protected public lands.

Theodore Roosevelt: the great conservationist

Theodore Roosevelt became one of the most powerful voices in the history of American conservation. An avid outdoorsman who cherished our nation's landscapes and wildlife, after becoming President in 1901, he used his executive power to establish 150 national forests, 51 bird reserves, four national game preserves, five national parks and 18 national monuments on over 230 million acres of public land.

Though he was born in New York City, and was sickly as a child, Roosevelt soon discovered a passion for the outdoors. His favorite activities were hiking, rowing, swimming, horseback riding, bird watching, hunting and taxidermy.

In 1903, President Theodore Roosevelt and naturist John Muir went on a camping trip to Yosemite, (which Lincoln had given to California). But then Teddy decided to add the area to the U.S. National Park system to afford it greater protection.

As President, Teddy Roosevelt went on a camping trip for several days exploring Yosemite Valley with famed naturalist John Muir. The two men discussed the importance of preserving the nation's natural resources for future generations to enjoy. So Roosevelt added Yosemite Valley and the Mariposa Grove to Yosemite National Park.

In a speech he delivered at the Grand Canyon on May 6, 1903, Roosevelt said "I want to ask you to keep this great wonder of nature as it now is. I hope you will not have a building of any kind, not a summer cottage, a hotel or anything else, to mar the wonderful grandeur, the sublimity, the great loneliness and beauty of the canyon. Leave it as it is. You cannot improve on it. The ages have been at work on it, and man can only mar it. What you can do is to keep it for your children, your children's children, and for all who come after you, as one of the great sights which every American if he can travel at all should see."

Years later, reflecting on this subject, Roosevelt commented, "There is a delight in the hardy life of the open. There are no words that can tell the hidden spirit of the wilderness that can reveal its mystery, its melancholy and its charm. The nation behaves well if it treats the natural resources as assets which it must turn over to the next generation increased and not impaired in value."

In 1904, Roosevelt created the U.S. Forestry Service to battle deforestation, and he even refused to have a Christmas Tree cut for the White House.

Hoover built a dam...and more

Though Republican President Herbert Hoover got the lion's share of the blame for not dealing effectively with the devastating effects of the Great Depression, he not only sought to promote American industries that would permit companies to achieve high profits but also pay high wages, and thereby allow employees to enjoy healthy recreation in the outdoors during their leisure hours. He expanded the Na-

Hoover dam still provides water and electricity to millions.

tional Park system by 40 percent.

Hoover's difficult negotiation—while serving as Commerce Secretary before becoming President—of the Colorado River Pact (involving an unprecedented seven states), permitted the construction of the dam that would bear his name and be finished after he left office.

Hoover Dam was an engineering marvel that helped control floods of fertile, but arid agricultural land, and provided a reliable source of water for millions of people in a multi-state area. It had a big impact on the development of Las Vegas. Herbert Hoover's vision was also to harness water to generate electricity, which the dam continues to do decades after it was built.

Located in the Black Canyon of the Colorado River, on the border between Nevada and Arizona, Hoover Dam is a concrete arch-gravity dam. It was constructed between 1931 and 1936 during the Great Depression. It took more than 10,000 people to build it, and at least 100 people died during its construction. (Originally known as Boulder Dam from 1933, it was officially renamed Hoover Dam, for President Herbert Hoover, by a joint resolution of Congress in 1947.)

Hoover Dam impounds Lake Mead, the largest reservoir in the United States by volume (when it is full). The dam's generators provide power for public and private utilities in Nevada, Arizona, and California. Hoover Dam is a major tourist attraction; nearly a million people tour the dam each year.

Clearly, the nation's 31st Chief Executive appreciated the outdoors, and one of his great passions was fly fishing. He would escape whenever he could to a summer weekend retreat he bought amongst the hemlocks on the eastern slope of the Blue Ridge Mountains in Virginia. The 164-acre property was on the site where two small streams merged to form the Rapidan River. Thirteen rustic cabins made up the compound and were nestled into a natural mountain setting enhanced by rock gardens, waterfalls and other stone structures. Here the Hoovers entertained family members, friends, Cabinet officers, and politicians for relaxing weekends of hiking, horseback riding, fishing, and conversation. It was about a three-hour drive from the city, but the fresh mountain air was a pleasant relief from Washington's heat and humidity; and the river offered excellent fishing, one of Hoover's favorite pastimes. The Hoovers paid for the land and the building supplies for the camp.

In a 1990 *New York Times* article, Robert M. Poole wrote, "Herbert Hoover was a keen angler, and the Rapidan was his favorite stream. Friends recall that he would come straight from the White

Still in a business suit, President Hoover dons waders for his favorite fishing spot.

House, so excited by the sight of the stream that he seldom took time to change clothes before sloshing in."

It can be said that Herbert Hoover combined his personal interest in conserving the environment with his personal qualifications as an engineer to find ways to responsibly use our natural resources.

Meanwhile, rising wheat prices in the 1920s and increased demand for wheat from Europe during World War I encouraged farmers to plow up millions of acres of native grassland to plant wheat, corn and other row crops. But as the United States entered the Great Depression, during the Republican administrations of Calvin Coolidge and then Herbert Hoover, wheat prices plummeted. Farmers tore up even more grassland in an attempt to harvest a bumper crop and break even. In retrospect, this was a great mistake.

Though historians regard GOP environmental policies of the 1920s as mistaken, ineffective and corrupt, Herbert Hoover—who served as President from 1929-33—pioneered some of the first broad environmental policies in the United States. The National Conference on Outdoor Recreation brought together wilderness advocates and urban planners, passage of the first federal law to limit oil pollution in navigable water, and began an ongoing effort to control the effects of industrialization. His advocacy of pleasant, affordable housing introduced the idea that the world we live in every day is our most important environmental concern.

FDR was one of our greatest environmental Presidents

Without a doubt, one of our nation's most courageous Presidents, Democrat Franklin Delano Roosevelt fought to overcome a serious physical impairment which would have discouraged a lesser person from pursuing such a taxing political career.

The land surrounding FDR's family home at Hyde Park, New York nurtured his life-long interest in nature and the environment. He first drew inspiration from this land as a young boy riding horseback with his father. Roosevelt's conservation ethic took root and flourished during a lifetime of exploring and caring for the place where, in 1912, he began to practice scientific forest management. He would plant half a million trees in Hyde Park during his lifetime—work conducted in cooperation with the State University of New York College of Science and Forestry at Syracuse.

His cousin, Theodore Roosevelt (TR), was an early influence. Franklin's political career, which mirrored TR's, also paralleled his maturing interest in conservation.

In August 1921, 39-year-old Franklin, at the time a practicing lawyer in New York, joined his family at their vacation home at Campobello, a Canadian island off the coast of Maine. As a former Assistant Secretary of the Navy, on August 5th, Roosevelt sailed up the New England coast with his friend and new employer, Van Lear Black, on Black's ocean-going yacht. On August 10th, after a day of strenuous activity, Roosevelt came down with a frightening illness characterized by fever, ascending paralysis, facial paralysis, prolonged bowel and bladder dysfunction, and numbness and hypersensitivity of the skin.

Most of the symptoms eventually resolved themselves, but he was left permanently paralyzed from the waist down. He was diagnosed with poliomyelitis at the time, but today doctors believe his symptoms were more consistent with Guillain–Barré syndrome (GBS)—an autoimmune neuropathy which Roosevelt's doctors didn't know was a diagnostic possibility.

FDR could only stand with the help of powerful leg braces, and even then usually had to hold onto somebody for balance, though he delivered many speeches by grasping a podium while standing—a real test of physical endurance. Although the public was vaguely aware of Roosevelt's disability, his optimistic personality overcame any notions that he had any serious limitations. News photographers respected his desire to project a healthy, vigorous persona, so avoided taking pictures of him on crutches or using a wheelchair.

In 1926, his belief in the benefits of hydrotherapy led him to found a rehab center at Warm Springs, Georgia. He later created the National Foundation for Infantile Paralysis, leading to the development of polio vaccines.

Despite his physical

A rare photo of FDR in a wheelchair.

limitations, upon assuming office as President of the United States, Franklin Roosevelt was a hard-working, industrious leader. Within his first month, he signed Executive Order 6101 (the Emergency Conservation Work Act), creating the Civilian Conservation Corps (CCC). This act addressed two pressing needs: unemployment and the repair of environmental damage, with one of the most successful New Deal programs.

Across the nation, the CCC planted three billion trees, built campgrounds and trails in more than 800 parks, removed invasive

Three million men—about five percent of the total U.S. male population—took part in the CCC during the agency's nine year history.

plants, improved wildlife habitats, and fought tree-killing insects and even forest fires. They also preserved historic sites, built roads, bridges, and dams. Forty million acres of farmlands benefited from erosion control projects, 154 million square yards of stream and lake shores were protected, 814,000 acres of range were revegetated, and 972 million fish were stocked.

When created in 1916, the National Park Service focused primarily on the conservation of spectacular landscapes, mostly in the West, and prehistoric native sites. FDR expanded the National Park Service mission in 1933 to include not only parks and monuments but also national cemeteries, national memorials, and national military parks. He also added the parks in Washington, D.C. The reorganization paved the way for inclusion of historic sites such as the Vanderbilt Mansion and Roosevelt's own home in New York state, which he made part of the National Park system in 1939 and 1943. With sweeping legislation, FDR was responsible for adding over one-quarter of the 411 areas in today's National Park Service system.

As he saw it, history, culture, and nature all played roles in the exceptional saga and unfolding legacy of the United States. "There is nothing so American," he said, "as our national parks."

As the National Park Service itself observes, "Franklin Roosevelt's New Deal programs, combined with his enthusiasm for conservation, laid a firm foundation for protecting the nation's natural bounty. The extent of the conservation projects carried on during the New Deal was far more reaching than anything attempted before. Soil erosion control, water conservation, the preservation of wildlife, and other environmental protection activities became a part of the everyday life and activities of American citizens. The importance of the work was new and inspiring. Under his leadership, FDR's programs introduced new concepts on a national level in planning for the responsible use of our natural and historic resources."

Franklin Roosevelt was nothing if not determined to reject failure, and he projected this onto the American people so that they too thought nothing was impossible. His indomitable spirit was displayed in many ways, even in private. Once, a visiting clergyman watched the paralyzed Roosevelt crawl from his desk across the floor to a shelf, then crawl back, clutching a book in his teeth. Asked why he had subjected himself to such an ordeal, FDR answered: "I felt I had to do it to show that I could."

He possessed a genius for reaching out to millions of people whose circumstances were utterly remote from his own experience. He conveyed a serenity, optimism and strength that touched hearts. It seemed entirely appropriate that a suffering nation should entrust its fate to a man who had also known suffering.

Among Roosevelt's many accomplishments was the establishment of the Shenandoah National Park in late 1935. It was 200,000

Skyline Drive in the Shenandoah National Park in Virginia.

acres of protected lands that include cascading waterfalls, spectacular vistas, and quiet wooded hollows that are a haven to deer, songbirds, and picturesque Skyline Drive. Unfortunately, to accomplish this, the state of Virginia and the government resorted in a lot of cases to "imminent domain," in which land was seized and landowners compensated in an amount that was often less than the true value and inflicted emotional harm. Many families and entire communities were forced to vacate portions of an area that became a National Park.

The same was true when Roosevelt dedicated the Great Smoky Mountains National Park in September 1940. As the National Park Service acknowledges, "Establishing most of the older parks located in the western United States, such as Yellowstone, was fairly easy. Congress merely carved them out of lands already owned by the government—often places where no one wanted to live anyway. But getting park land in [the east] was a different story. The land that became Great Smoky Mountains National Park was owned by hundreds of small farmers and a handful of large timber and paper companies.

"Surprisingly, motorists had the biggest role in the push for a National Park. The newly formed auto clubs, mostly branches of the Automobile Association of America (AAA), were interested in good roads through beautiful scenery on which they could drive their shiny new cars."

However, since the federal government was not allowed to buy land for National Park use, political donations and individual state

FDR speaking at the dedication ceremonies for the Great Smokies at the Rockefeller Memorial at Newfound Gap astride the Tennessee - North Carolina state line.

funds were relied on to acquire the land. This was a tedious process that had begun back in Calvin Coolidge's day but only reached completion during Roosevelt's term.

Aside from his interest in preserving the country's natural wonders, Franklin Roosevelt preached that Americans must pay heed to the imperatives of living in harmony with the land. He said that "Men and nature must work hand in hand. The throwing out of balance of the resources of nature throws out of balance also the lives of men."

No truer words were ever spoken. One major consequence of the public's previous disregard for respecting nature culminated in an unprecedented natural disaster.

What was termed "the dust bowl" vividly illustrates how man's failure to understand his environment can have disastrous consequences. The origins of the problem can be traced back to the period after the Civil War, when a series of federal land acts coaxed pioneers westward by incentivizing farming in the Great Plains.

Severe drought hit the Midwest and Southern Great Plains in 1930. That year, weather patterns shifted over the Atlantic and Pacific oceans. The Pacific grew cooler than normal and the Atlantic became warmer. Kimberly Amadeo explains, "The combination weakened and changed the direction of the jet stream. That air current carries moisture from the Gulf of Mexico up toward the Great Plains. It then dumps rain when it reaches the Rockies. When the jet stream moved south, rain never reached the Great Plains.

"Tall prairie grass once protected the topsoil of the Midwest. But once farmers settled the prairies, they plowed over 5.2 million acres of the deep-rooted grass. Years of over-cultivation meant the soil lost its richness. When the drought killed off the crops, high winds blew the remaining topsoil away. Parts of the Midwest still have not recovered.

"As the dust storms grew, they intensified the drought. The airborne dust particles reflected some sunlight back into space before it could reach the earth. As a result, the land cooled. As temperatures dipped, so did evaporation. The clouds never got enough moisture to release rain."

Severe dust storms, often called "black blizzards"—which swept through the Great Plains states of Texas, Oklahoma, Kansas and Nebraska but ultimately affected 23 states in the Mississippi and Ohio river valleys—carried topsoil as far east as Washington, D.C. and New York City, and even coated ships in the Atlantic Ocean with dust.

Clouds of dust swept across the land, due to bad farming practices.

Billowing clouds of dust would darken the sky, sometimes for days at a time. In many places, the dust drifted like snow and residents had to clear it with shovels.

John Steinbeck wrote in *The Grapes of Wrath*, "Houses were shut tight, and cloth wedged around doors and windows, but the dust came in so thinly that it could not be seen in the air, and it settled like pollen on the chairs and tables, on the dishes."

Dust worked its way through the cracks of even well-sealed homes, leaving a coating on food, skin and furniture. People developed respiratory ailments and thousands died

On May 11, 1934, a massive dust storm two miles high traveled 2,000 miles to the East Coast, blotting out monuments such as the Statue of Liberty and the U.S. Capitol.

The History channel reports the "worst dust storm occurred on April 14, 1935. News reports called the event Black Sunday. A wall of blowing sand and dust started in the Oklahoma Panhandle and spread east. As many as three million tons of topsoil are estimated to have blown off the Great Plains during Black Sunday."

As high winds and choking dust swept the plains states, people and livestock were killed and crops failed across the entire region. "The Dust Bowl intensified the crushing economic impacts of the Great Depression and drove many farming families on a desperate migration in search of work and better living conditions," the History Channel concludes.

By 1934, an estimated 35 million acres of formerly cultivated land had been rendered useless for farming, while another 125 million acres—an area roughly three-quarters the size of Texas—was rapidly losing its topsoil.

President Franklin D. Roosevelt established a number of measures to help alleviate the plight of poor and displaced farmers. He also addressed the environmental degradation that had led to the Dust Bowl in the first place. He commented, "A nation that destroys its soils destroys itself."

Congress established the Soil Erosion Service and the Prairie States Forestry Project in 1935. These programs put local farmers to work planting trees as windbreaks on farms across the Great Plains. The Soil Erosion Service, now called the Natural Resources Conservation Service (NRCS) implemented new farming techniques to combat the problem of soil erosion.

Regular rainfall returned to the region by the end of 1939, bringing the Dust Bowl years to a close. The economic effects, however, persisted. Roughly 2.5 million people left the Dust Bowl states during the 1930s. Population declines in the worst-hit counties—where the agricultural value of the land failed to recover—continued well into the 1950s.

FDR's time in office, of course, was not dedicated solely to enhancing the environment or dealing with the dust bowl. The attack by the Japanese on Pearl Harbor on December 7, 1941 drew the United States into a World War, and it transformed American society which had to instantly re-tool to get on a survival footing. WWII required the commitment of every man, woman and child in America.

One interesting postscript to Roosevelt's time in office (he was elected, but did not live to fulfill, four consecutive terms), was that the President would sometimes slip away to commune with nature.

Historian David Brinkley, who wrote a book called *Rightful Heritage: Franklin D. Roosevelt and the Land of America*, revealed the little known fact that FDR once "went to the Galapagos Islands in 1938—on a scientific expedition in conjunction with the Smithsonian—and nobody had ever written about it, because there was no journal, no record. The President just disappeared for weeks! Theodore Roosevelt wrote all the time, he wrote over 35 books and 100,000 letters; whenever he went on an outdoor adventure, he would write about it and document it, where FDR did not do that." Fortunately, someone who went on the trip with Roosevelt kept a diary, which Brinkley was able to track down.

The National Park Service affirms "[President Franklin Roosevelt] left a conservation legacy essential to a healthy 21st-century environment—one that we continue to build upon today."

Truman liked the Florida Keys

FDR's successor, Harry S Truman, was understandably preoccupied with the end of WWII. In November 1946, after he had finished 19 months in office, he was physically exhausted. His doctor ordered a warm vacation. To escape the winter cold of Washington, D.C., Truman went down to Key West, Florida, where he took refuge in a single family house on a submarine base there, that was originally

Wearing an old hat and a sport shirt with the tails worn outside his pants, President Truman takes a stroll along the beach at Key West, Florida in December 1947. Fort Taylor (in the background) was used during the Civil War. Truman also dedicated the Everglades National Park, with the ceremony broadcast on nationwide radio.

Everglades National Park – Tarpon Bay, looking southwest towards Florida Bay.

intended to house the base commandant. Truman enjoyed his time there so much, the residence came to be known as the Winter White House, and the President spent 175 days there during 11 visits.

Whenever Truman took up residence in the Florida Keys, Cabinet members and foreign officials were regular visitors for fishing trips and poker games. Presidents up to Bill Clinton also occasionally used the house as a retreat or for official functions.

Truman came to appreciate the importance of the state's natural habitat. So about a year after his first visit, he dedicated the Florida Everglades—a 1.5-million-acre wetlands preserve on the southern tip. Located at the confluence of temperate North America and the tropical Caribbean, Everglades National Park is home to representative flora from both climes. The optimal growing conditions that are prevalent throughout south Florida foster a lush growth of plant life that sustains a diverse complex of flora. The Everglades serve as important habitat for a number of endemic and legally protected species.

Often compared to a grassy, slow-moving river, the Everglades is made up of coastal mangroves, sawgrass marshes and pine flatwoods that are home to hundreds of animal species. Among the Everglades' abundant wildlife are the endangered leatherback turtle, Florida panther and West Indian manatee. Besides alligators, the crocodile also populates the Everglades. The area is an international treasure as well—a World Heritage Site, International Biosphere Reserve, and a

Wetland of International Importance.

At the dedication ceremony, Harry Truman observed, "Here is land, tranquil in its quiet beauty, serving not as the source of water, but as the last receiver of it. To its natural abundance we owe the spectacular plant and animal life that distinguishes this place from all others in our country."

The next year, President Truman signed the Water Pollution Control Act of 1948, which was the nation's first water-quality law.

As Alonzo L. Hamby, a Professor of History at Ohio University writes, "Truman protected the New Deal and—with a rise in the minimum wage in 1949 and the enlargement of Social Security in 1950—built upon its achievements. He pushed forward the cause of African-American civil rights by desegregating the military, by banning discrimination in the civil service, and by commissioning a federal report on civil rights. Just as important, Truman spoke out publicly on the matter."

His handling of the Korean war and scandals in his administration, plus the red-baiting of Joe McCarthy all made Truman's years in the White House difficult. But after he left office, his public image was significantly restored as Americans began to see him, as "a feisty everyman from 'Middle America' rather than a partisan Washington, D.C., politico," Hamby concludes.

Eisenhower rebuilds a post World War II America

During the subsequent administration of Republican Dwight Eisenhower, when Alaska became a state, the President realized that there were vast areas of land that needed to be allocated among the state and federal governments, Eskimos and other Native Americans, and some private interests. The House of Representatives passed legislation recommending that nine million acres of pristine land along the northern shore of Alaska be protected permanently from any commercial development. But two Alaska Senators blocked Senate action. So in 1960, Eisenhower accomplished his goal by establishing the Arctic National Wildlife Range "for the purpose of preserving unique wildlife, wilderness and recreational values."

He had the wisdom and foresight to include the entire ecosystem, both north and south of the Brooks Range, including the biologically rich Coastal Plain, which is essential to the integrity of this ecosystem. The Coastal Plain is the heart of this wild Arctic ecosystem, supporting the 197,000-animal Porcupine Caribou Herd, millions of migratory birds, and a full-complement of large predators, such as wolves, grizzly bears, and polar bears.

In 1980, Congress enlarged the original range to protect additional wildlife habitat and to establish the Arctic National Wildlife Refuge. In the same move, Congress closed the Coastal Plain to oil development, and any move to allow oil drilling activity would require a new Act of Congress.

(President Donald Trump recently signed a tax bill into law that includes a provision that opens the Arctic National Wildlife Refuge to oil and gas drilling to offset massive corporate tax cuts but as of this writing, a federal judge blocked Trump's action. Its fate is uncertain).

The Arctic National Wildlife Refuge is an iconic American treasure. Birds migrate from across the United States and from six continents in order to feed and reproduce in the Arctic Refuge, taking advantage of the burst of plant and insect life during the long days of the Arctic summer.

Also during the Eisenhower years, legislation was passed to provide more funds for acquiring wildlife refuges and to make the improvement of fish and wildlife resources a specific purpose in federal water resource construction, plus legislation long urged by conservationists was enacted to make applicable on military lands the fishing, hunting, and trapping laws of the states and territories in which such lands are located. A commission was established to survey the future

Polar bears could possibly go extinct in our lifetimes. Only 30,000 or so bears exist today, and roughly 50 bears come into the Arctic Refuge each year in September.

President Dwight Eisenhower's Interstate Highway System has had a major impact on air pollution in the United States.

needs of our expanding population for recreational areas. The President asserted, "Part of our Nation's precious heritage is its natural resources," and he was determined to protect them.

Eisenhower later reflected, "Those of us who venerated Theodore Roosevelt's example were determined that, with our rapidly increasing population and proliferating industrialization, our extraordinary natural recourses and national beauty would not be 'civilized off the face of the earth.'"

Probably the one action that Dwight Eisenhower took that had the biggest impact on the environment was his signing in 1956 of the bill that created the Interstate Highway System (IHS) across the nation. It became the largest public works project in the history of the world (only later surpassed by China's Three Gorges Dam).

With millions more Americans driving on this expansive highway grid (that was not totally completed for 35 years), this obviously has inflicted major harm on our country.

As Winslow Johnson of the Swarthmore College Environmental Studies program notes, "The environmental costs of this increase in driving come in the form of higher emissions of air pollutants, CO_2 and other greenhouse gasses, and greater exposure to lead, until its 1986 phasing out as an additive in gasoline. Particularly harmful was the 70 percent increase in driving during the 1960s and 1970s, before the federally mandated use of catalytic converters [on vehicles] to reduce emissions and the implementation of higher mileage standards to lower overall gas consumption."

Air pollution was becoming so bad from vehicles, that there were numerous deadly smog attacks over several years that killed hundreds and sickened thousands more in Los Angeles, New York, and other North American cities.

The IHS also redistributed Americans and changed the places where they live. The environmental impact of an increase in population density is obvious: an increase in vehicle emissions, water usage and runoff, and waste coupled with a decrease in open space and other wild habitat. Therefore, as the IHS has had an impact on the distribution of the American population, it has also created new pockets of environmental contamination and destruction.

As Meir Rinde of the Science History organization observes, "After World War II the scope of Americans' conservation concerns broadened as they began

President Dwight Eisenhower

worrying about the increasingly visible despoliation of the environment. The baby boom and economic prosperity spurred rapid suburbanization, eating up farmland and open space. Smoke-belching industrial plants proliferated to process raw materials and supply families with mass-produced goods—cars, refrigerators, stereos, air conditioners, toys, processed foods—along with electricity to illuminate their ranch homes and gasoline to power their Chevys and Fords.

"At the same time, paid vacation, shorter work weeks, labor-saving devices like washing machines, and a vastly expanded highway system made it easier for suburbanites to visit national parks and to camp and fish. More of them began to see unspoiled wilderness as a picturesque escape and spiritual salve rather than a mere storehouse of lumber and minerals."

Kennedy Explores "The New Frontier"

The passing of the torch from one generation to another, as John Fitzgerald Kennedy became President of the United States in 1961 also represented the embrace of new ideas, including how to deal with an increasing population and a decreasing number of natural resources.

Growing up in the Boston, Massachusetts area, Kennedy had an urbanite's perspective. Still, from his earliest days, JFK developed an affinity for the sea, nurtured when his family—which included nine children—escaped the city and spent summers at the expansive three-home compound patriarch Joseph Kennedy had acquired on Cape Cod along Nantucket Sound in Hyannis Port, MA. The main house commands sweeping views of the ocean from its long porches. It was here that John acquired a lifelong interest in sailing and other competitive activities.

When he was 15 years old, he was gifted with a 25-foot Wianno Senior, a classic wooden gaff-rigged sloop made nearby on Cape Cod. JFK named her *Victura*, Latin for "about to conquer"—fitting for a young man with big dreams. He went on to other, larger boats—most notably the Sparkman & Stephens designed *S/Y Manitou*, and the Presidential motor yacht *Honey Fitz*, but it was *Victura* that captured and held his imagination and his heart. Friends said that he often told them he was never happier than at the helm of *Victura*. In fact, he even doo-

A young Jack Kennedy getting ready to set sail aboard the *Victura*.

dled a sketch of it on hotel stationery just a few hours before he was slain in Dallas, Texas—a paper that fortunately was retrieved by a hotel maid from the waste basket in the room the President had just vacated.

During his lifetime, Kennedy had been reinvigorated by walks on the beach at his family's compound at Hyannis Port and his swims in the ocean near their home at Palm Beach, Florida.

Kennedy reflected, "I really don't know why it is that all of us are so committed to the sea, except I think it is because in addition to the fact that the sea changes, and the light changes, and ships change, it is because we all came from the sea. And it is an interesting biological fact that all of us have in our veins the exact same percentage of salt in our blood that exists in the ocean, and, therefore, we have salt in our blood, in our sweat, in our tears. We are tied to the ocean. And when we go back to the sea, whether it is to sail or to watch it we are going back from whence we came."

Despite being born to wealth, JFK commented in his typically egalitarian way that "I do not know why it should be that six or seven percent only of the Atlantic coast should be in the public sphere and the rest owned by private citizens and denied to millions of our fellow citizens." He wanted the ocean and beaches accessible to everyone.

As President, he designated the nation's first three national seashores—on Massachusetts' Cape Cod, Padre Island in South Texas, and Point Reyes, just north of San Francisco. They were not only selected to represent the three major bodies of water that shaped the country's continental coastline (the Atlantic, Gulf of Mexico, and the Pacific), they were within several hours' drive of major population centers (the northeast corridor; Houston and San Antonio; and the Bay Area).

Kennedy's administration sought and secured Congressional funding through the Housing Act of 1961 for advanced water-quality projects, sewage-treatment facilities, and urban parklands.

The President presciently warned in 1962, "We are going to have 300 million people by the end of this century, and we have to begin to make provisions for them. We do not want, for example, this eastern coast to be one gigantic metropolitan area stretching from north of Boston to Jacksonville, Florida, without adequate resources for our people."

Also that year, marine biologist Rachel Carson published her epic book, *Silent Spring*, documenting the adverse environmental effects caused by the indiscriminate use of pesticides. Carson accused the chemical industry of spreading disinformation, and public officials of accepting the industry's marketing claims unquestioningly.

Ms. Carson became interested in the subject several years earlier when a friend wrote a letter to the editor of *The Boston Herald* describing the death of birds around her property resulting from the aerial spraying of DDT to kill mosquitoes. During four years of research, Carson discovered a sizable community of scientists who were documenting the physiological and environmental effects of pesticides on humans and wildlife. She took advantage of her personal connections with many government scientists, who supplied her with confidential information on the subject.

During the course of her research, Rachel Carson became bedridden for months, underwent a mastectomy, then discovered her cancer had metastasized. Still, she persisted. Carson attended the White House Conference on Conservation in May 1962 and even sent a proof copy of her impending book to Supreme Court Justice William O. Douglas, a long-time environmental advocate who supplied her with information and later even publicly endorsed her book.

Just before the publication of *Silent Spring*, the story of the birth defect-causing drug thalidomide broke, inviting comparisons between Carson and Frances Oldham Kelsey, the Food and Drug Administration reviewer who had blocked the drug's sale in the United States. The *New Yorker* and *Atlantic* magazine ran excerpts of Rachel's book, and it was chosen as a "Book Of the Month" club selection.

Her work was met with fierce opposition by chemical companies. Robert White-Stevens, a former Cyanamid chemist, scoffed, "If man were to follow the teachings of Miss Carson, we would return to the Dark Ages, and the insects and diseases and vermin would once again inherit the earth." But the author made it clear she was not advocating the elimination of all pesticides, but was instead encouraging responsible and carefully managed use with an awareness of the chemicals' impact on ecosystems.

After ten to fifteen million Americans viewed Rachel Carson on a *CBS Reports* television special, her book and its thesis soared. In one of her last public appearances, Carson testified before President John F. Kennedy's Science Advisory Committee, which issued its report on May 15, 1963, largely backing Carson's scientific claims.

Carson's work had a powerful impact on the environmental movement. *Silent Spring* became a rallying point for the new social movement in the 1960s. According to environmental engineer and Carson scholar H. Patricia Hynes, "*Silent Spring* altered the balance of power in the world. No one since would be able to sell pollution as the necessary underside of progress so easily or uncritically."

Carson's book spurred a reversal in the United States' national pesticide policy, led to a nationwide ban on DDT for agricultural uses, and helped to inspire an environmental movement that ultimately led to the creation of the U.S. Environmental Protection Agency, which describes itself as no less than "the extended shadow of Rachel Carson." Ms. Carson was posthumously awarded a Presidential Medal of Freedom.*

A milestone of John F. Kennedy's Presidency—which had unexpected benefits to the environment as well as humans in general—was when he announced the goal of "landing a man on the moon, before this decade is out, and returning him safely to earth."

Kennedy characterized space as a new frontier, invoking the pioneer spirit that dominated our nation's folklore. He infused the speech with a sense of urgency and destiny, and emphasized the freedom enjoyed by Americans to choose their destiny rather than have it chosen for them. Although he called for competition with the Soviet Union, he also proposed making the Moon landing a joint project.

In a stirring speech delivered to 40,000 people gathered in Rice University stadium on a sweltering day in September 1962, JFK declared "We choose to go to the moon in this decade and do the other things, not because they are easy, but because they are hard, because that goal will serve to organize and measure the best of our energies and skills, because that challenge is one that we are willing to accept, one we are unwilling to postpone, and one which we intend to win, and the others, too." A bona fide war hero, JFK spoke with authority.

The National Aeronautics and Space Administration (NASA),

---

*Years later, the *Wall Street Journal*, the *National Review, Reason* and of course FOX News, all claimed that Rachel Carson was responsible for more deaths worldwide than Hitler. Leading the charge against Carson is the Competitive Enterprise Institute, a Washington think tank "dedicated to advancing the principles of free enterprise and limited government." It claims that Carson's criticism of DDT left millions vulnerable to malaria, with 100 million people—mostly women and children—dying since 1972, the year DDT was supposedly banned. But here again, rightwing elements are distorting the truth. Carson never sought the total ban on DDT—only that its use be carefully controlled—and use of the pesticide is still allowed for combatting insect-borne diseases. According to the *New Scientist*, each year about 1,000 tons of DDT are still released worldwide. Carson had presciently warned that over-use made mosquitoes resistant. In fact, by 1972, 19 species of mosquitoes thought to transmit malaria were resistant to DDT. There are now better alternatives to the pesticide. *The Christian Science Monitor* reports that "In 1991, Vietnam switched from a DDT-based campaign to one focusing on rapid treatment, mosquito nets, and a different type of insecticide. The World Health Organization reports that malaria fatalities dropped by 97 percent."

along with government and private research, have been responsible for many advancements that benefit our lives on earth.

NASA administrator Michael Griffin later said "We see the transformative effects of the Space Economy all around us through numerous technologies and life-saving capabilities. We see the Space Economy in the lives saved when advanced breast cancer screening catches tumors in time for treatment, or when a heart defibrillator restores the proper rhythm of a patient's heart... We see it when weather satellites warn us of coming hurricanes, or when satellites provide information critical to understanding our environment and the effects of climate change... Technologies developed for exploring space are being used to increase crop yields and to search for good fishing regions at sea."

JFK at Rice University.

The myriad ways in which the U.S. space program has improved life on earth are virtually incalculable.

In September 1963, two months before he was assassinated, John Kennedy embarked on a five day, 11-state, 15-speech coast-to-coast trip that would take him from Pennsylvania to Wisconsin, Minnesota, North Dakota, Wyoming, Montana, Utah, Nevada, California, Oregon and Washington, to raise national awareness of the challenges confronting our environment. The President noted, "we are reaching the limits of our fundamental needs of water to drink, of fresh air to breathe, of open space to enjoy, of abundant sources of energy to make life easier." To respond to these pressures facing urban and rural America would, he said, require the creation of new ideas, "the embrace of disciplines unknown in the past."

On his multi-state tour, JFK formally dedicated the Pinchot Institute for Conservation Studies, whose mission was to research and evaluate the environmental pressures then challenging the country and to offer remedies through conservation-education programs, thus building the consensus for social change. Thereby Kennedy laid the groundwork for Lyndon Johnson's Great Society, with its emphasis on public health, social justice, and environmental protection.

Declaring that he was in favor of a "a third wave of conservation in the United States following that of Theodore and Franklin Roosevelt," President Kennedy promoted the passage of a land and water conservation fund then bottled up in Congress—funds from which the government would purchase wetlands, wildlands, and other threatened terrain.

There's an amusing anecdote that offers insight into John Kennedy's interest in the natural world. On his late September 1963, trip, the President stayed one night in a cabin in Lassen Volcanic National Park in California. The next morning, he spotted a deer from his cabin window, so he went outside still clad in his pajamas and red bedroom slippers, and began feeding it bread. After going back inside and sitting down for breakfast, he asked why there was no toast with his eggs, and was informed he had fed all the bread to the deer!

President Kennedy, still clad in his pajamas and red bedroom slippers, feeds a deer.

Lyndon Johnson turned off White House lights

Delivering his first State of the Union address to Congress in January 1964 following the death of John Kennedy, President Lyndon Johnson included a passage in his remarks in which he stated: "for over three centuries the beauty of America has sustained our spirit and has enlarged our vision. We must act now to protect this heritage. We must make a massive effort to save the countryside and to establish—as a green legacy for tomorrow—more large and small parks, more seashores and open spaces than have been created during any other period in our national history. A new and substantial effort must be made to landscape highways to provide places of relaxation and recreation wherever our roads run. Within our cities imaginative programs are needed to landscape streets and to transform open areas into places of beauty and recreation. We will seek legal power to prevent pollution of our air and water before it happens. We will step up our effort to control harmful wastes, giving first priority to the cleanup of our most contaminated rivers. We will increase research to learn much more about the control of pollution."

Mr. Johnson put his conservation ethic into daily practice. The President was known for wandering the White House, turning off lights in rooms he thought were empty, sometimes to the surprise of people working in those rooms!

Intriguingly, Lyndon Johnson, already knew of the dangers of climate change and spoke of them in a special message to Congress shortly after his 1965 inauguration. "Air pollution is no longer confined to isolated places," he said. "This generation has altered the composition of the atmosphere on a global scale through radioactive materials and a steady increase in carbon dioxide from the burning of fossil fuels."

Moreover, a report written by his science advisory committee later that year confirmed the climate threat, describing atmospheric carbon dioxide from fossil fuels as "the invisible pollutant," and foretelling many of the effects of today, including the melting of the Antarctic ice cap, increasing ocean acidity, and sea level rise.

"The climate changes that may be produced by the increased CO2 content could be deleterious from the point of view of human beings," the report warned.

Yet think of how obstreperous some rightwing politicians still are about accepting the science behind climate change this many years later.

Lady Bird Johnson's counterpart to her husband's interest in

President Lyndon Johnson and the First Lady enjoying a field of flowers.

conservation was to help make America look better. At her urging, Congress passed the Highway Beautification Act in 1965 that called for control of unsightly billboards which had proliferated along the nation's growing Interstate Highway System. It also required certain junkyards along Interstate or primary highways to be removed or screened and encouraged scenic enhancement and roadside development. Mrs. Johnson delighted in seeing masses of wildflowers planted along roadways across the nation.

"Ugliness is so grim," Lady Bird once said. "A little beauty, something that is lovely, I think, can help create harmony which will lessen tensions."

That belief—that beauty can improve the mental health of a society—and her determination to make the United States a more beautiful place became the First Lady's legacy. Throughout her time in the White House, she fought to make American cities more appealing by planting flowers or adding park benches.

Adrian Benepe, former Commissioner of Parks & Recreation in New York City, explains that Mrs. Johnson "spent much of her childhood in the meadows and woodlands of Karnack, Texas... There is a lot of speculation as to why President and Lady Bird Johnson were so keenly interested in the environment and natural beauty; some think it is rooted in Mrs. Johnson's loss of her mother at a very young age, after which she found solace in the flowers and plants around her childhood home. President Johnson—who led the passage of groundbreaking civil rights legislation and many other significant domestic policy acts of the 'Great Society'—fully acknowledged his wife's role as instigator of, and inspiration and advocate for much of his environmental legislation."

Mrs. Johnson's special emphasis was on making Washington, D.C., look better. She wanted the nation's capitol cleaned up and enlisted the help of volunteers and philanthropists who worked to rid the city of blighted areas. As a result of her efforts, D.C. gained hundreds of landscaped parks and thousands of daffodils, azaleas and dogwood trees during Lady Bird's tenure that endure to this day.

Mrs. Johnson saw her beautification projects as helping soothe the nation at a time when the Vietnam War, civil rights and other highly charged political topics fomented division. Lady Bird believed that a cleaner, more beautiful country could calm people and bring them together. She saw her conservation and beautification work as interwoven with President Johnson's Great Society agenda.

She took such delight in a sweep of uncluttered hills and swirling clean rivers, well-planned urban development and her own plantings.

Despite the fact that Lyndon Johnson presided over a horrific war in Southeast Asia and many people died, it is interesting that he commented in one speech that "to sustain an environment suitable for

The Johnsons enjoying a moment of solitude at their Texas ranch.

man, we must fight on a thousand battlefields. Despite all of our wealth and knowledge, we cannot create a redwood forest, a wild river, or a gleaming seashore."

Adrian Benepe points out that "At the conclusion of the Johnson administration in 1968, the President presented his wife with a plaque adorned with 50 pens used to sign 50 laws related to natural beauty and conservation, and inscribed: 'To Lady Bird, who has inspired me and millions of Americans to try to preserve our land and beautify our nation. With Love from Lyndon.'"

Benepe asks rhetorically, "So what would President and Mrs. Johnson think now, as partisan politics and fringe political movements work to strip environmental legislation of its power, to sell off federal lands for profit and exploitation...?"

As an environmentalist, Richard Nixon (surprisingly) ranks #3

LBJ's successor, Republican Richard Nixon, was not only the 20th century's greatest master of geopolitics, he surprisingly is ranked by environmentalists as the third greatest President on environmental issues.

On January 29, 1969—only nine days after Nixon took office— a freshly drilled oil well off the coast of Santa Barbara, California exploded, and oil began leaking into the ocean at the rate of 210,000 gallons a day. It befouled the nearby beaches and coated wildlife in black tar. The spill "shocked Americans," writes historian J. Brooks Flippen, "placing environmental protection on the front burner in a way it never had been before, turning a concerned public into an activist one."

President Nixon visited the area, strolling the beach. He commented to reporters, "What is involved is something much bigger than Santa Barbara. What is involved is the use of our resources of the sea and the land in a more effective way, and with more concern for preserving the beauty and the natural resources that are so important to any kind of society that we want for the future. I don't think we have paid enough attention to this... We are going to do a better job than we have done in the past." He kept his promise.

Soon after the Santa Barbara oil spill came a famous *Time* magazine cover showing Ohio's oozing, intensely polluted Cuyahoga River on fire.

The Science History organization reports that "By the late 1960s the burgeoning middle class was discovering that environmental problems threatened their children, their property values, and their

When he was a young Congressman, the Nixons rode their bikes to view the cherry blossoms in Washington but returned as President and First Lady in April 1969.

lifestyles." Journalist Philip Shabecoff wrote that they worried "about PCBs in mother's milk, about polybrominated biphenyls in Michigan cattle, about poisons leaking from rusty drums in their backyards, about strontium 90 from atmospheric testing of nuclear weapons."

By the time Nixon was elected, the nation was pumping out 200 million tons of air pollutants and throwing out 100 million automobile tires and 30 billion glass bottles annually, with much of the refuse piled in mountainous open dumps.

The President had picked as his top domestic adviser John Ehrlichman, who believed in safeguarding natural resources. In turn,

Ehrlichman was guided by Democratic savant Daniel Patrick Moynihan, who sent him a memo in which he explained, "Carbon dioxide in the atmosphere has the effect of a pane of glass in a greenhouse. The CO2 content is normally in a stable cycle, but recently man has begun to introduce instability through the burning of fossil fuels. At the turn of the century several persons raised the question whether this would change the temperature of the atmosphere. Over the years the hypothesis has been refined, and more evidence has come along to support it. It is now pretty clearly agreed that the CO2 content will rise 25% by 2000. This could increase the average temperature near the earth's surface by 7 degrees Fahrenheit. This in turn could raise the level of the sea by ten feet. Goodbye New York. Goodbye Washington, for that matter. We have no data on Seattle."

Over the next few years, President Nixon* went on to propose an ambitious and expensive pollution-fighting agenda to Congress, created the Environmental Protection Agency (EPA), the National Oceanic and Atmospheric Administration (NOAA), and other key ele-

---

*Though our 37th President has long been unfairly caricatured by many as an evil, corrupt and ruthless politician, the truth is Nixon was one of our most cerebral Presidents with a deep knowledge of and respect for world history that underpinned his impressive strategic planning. He was a man who not only possessed great political acumen, but impressive analytical skills, not just in foreign policy but also in the domestic realm as well. He was not a rigid ideologue and despite being the first President in 120 years, since Zachary Taylor, to face a Congress controlled in both houses by members of the opposition party, he accomplished much.

Nixon came to the aid of Israel with a much-needed arms airlift during the 1973 Yom Kippur War—an astonishing 567 deliveries—and he was close to Golda Meir, the Israeli prime minister who toasted him; Moshe Dayan, the defense minister; and Yitzhak Rabin, the ambassador to the United States. Rabin would break diplomatic protocol and campaign for Nixon in 1972. President Nixon signed the ABM Treaty and interim SALT agreement with the Soviet Union to limit nuclear weapons. He brilliantly initiated a rapprochement with China. He lowered the voting age to 18, created the all-volunteer military, launched the war on cancer as well as hunger, imposed wage and price controls to help consumers, expanded enforcement of affirmative action and the more thorough desegregation of Southern schools, and signed Title IX that bans discrimination based on sex in programs that receive federal funding. Nixon used the "peace dividend" from reducing troops in Vietnam to finance social welfare services and enforce civil rights through the Equal Employment Opportunity Commission. As a result, from 1970 to 1975, spending on human resource services exceeded spending for defense for the first time since World War II.

Nixon was also one of the most qualified men to serve in the Presidency. During World War II he was assigned to Marine Aircraft Group 25 and the South Pacific Combat Air Transport Command (SCAT), supporting the logistics of operations in the South Pacific Theater. He received multiple military awards before being elected to the House of Representatives twice, served as a Senator for one term, then been Vice President under Dwight Eisenhower for eight years.

ments of the nation's modern environmental infrastructure. Nixon signed the Clean Air Act of 1970 into law, which gave the EPA the authority to regulate air quality. Over the next half-century that measure and further amendments would help reduce by nearly 70% the total emissions of six major pollutants—carbon monoxide, lead, ground-level ozone, nitrogen dioxide, particulate matter, and sulfur dioxide—even as the U.S. population continued to climb and the country's economy expanded.

President Nixon also established a White House office on energy policy and created the interdepartmental position of energy "czar" to coordinate it, eventually leading to a system in which officials could allocate resources on a regional basis to ensure that New England didn't run out of gas while service stations in West Texas were overstocked.

Later, Nixon approved the Clean Water Act, the Pesticide Control Act, the Marine Mammal Protection Act, and the Endangered Species Act in 1973, which strengthened earlier protections for endangered species initiated during the Johnson administration.

Nixon also made other moves environmentalists favored, such as permanently stopping construction of the controversial Cross

A very formal man, who was seldom photographed without wearing a business suit, Richard Nixon enjoys a cruise off Maine with his daughter, Julie in 1971

Florida Barge Canal, which had already sliced partway across the Florida peninsula and would have decimated wildlife in the Ocklawaha River ecosystem. In his second environmental address he proposed greater EPA authority over pesticide regulation, more money for sewage-treatment centers, and funding for states to develop environmentally friendly land-use programs. Nixon surprised everyone.

As Meir Rinde of the Science History organization notes, "He had visited Yosemite National Park and the Grand Canyon as a child and 'would lighten up' at mention of parks, according to John Whitaker, his environmental-policy aide. As far back as 1962 Ehrlichman had tutored Nixon on fundamentals of the issue, beginning during a boat trip they took on the Puget Sound. Nixon was also close to his brother Ed, who had studied geology and had environmentalist leanings."

Stephen Hess, a senior fellow emeritus in the Governance Studies program at the Brookings Institution opines "He was a very environmentally aggressive President, although that's not what people tend to remember about him."

Historian Paul Charles Milazzo points to Nixon's creation of the EPA, and to his enforcement of 1973's Endangered Species Act, stating that "though unsought and unacknowledged, Richard Nixon's environmental legacy is secure."

There was unfortunately a Ford in our future

Gerald Ford's brief term in office after Nixon resigned in 1974 was marked by his disinterest in nature. Though an avid skier on the slopes of Vail, Colorado, he was never known as a strong advocate of environmental regulation. In fact, Ford supported measures to relax those regs, specifically in the areas of auto emission controls and regulation of strip mining.

Critics cite his failure to acknowledge the validity of the EPA as an independent agency as an important difference between him and Nixon. Where Nixon appeared to sincerely be concerned with preservation of the environment and natural resources, Ford saw environmental regulation as another bureaucratic hurdle over which his administration would have to jump. Such indifference would eventually mark Ford's environmental legacy as poor. The National League of Conservation Voters said that when it came to environmental issues, Gerald Ford was "hopeless."

Mr. Ford exhibited a similar disinterest in the truth when he served on the Warren Commission looking into John Kennedy's death.

Americans resisted giving up their gas-guzzling vehicles, even in the face of oil shortages and rising fuel prices, expecting the government to solve the problems.

Jimmy Carter installs solar panels on the White House roof

Jimmy Carter was commonly known as a humble man and a peanut farmer from Georgia who surprised everyone but himself by coming from political obscurity and winning the Presidency from Gerald Ford.

As a young man, Carter joined the Navy ROTC program at the Georgia Institute of Technology and his high grades enabled him to enter the United States Naval Academy, after which he began his Navy career as a nuclear engineer.

It was during his time in the Navy that Carter had his first experience with a near environmental disaster when an atomic reactor began to melt down in 1952 at Chalk River, Canada. He and his Naval unit were dispatched to help with the reactor and the grim reality of how close the reactor and radiation came to endangering the surrounding area had a lasting influence on Carter. The Chalk River accident shaped his thinking about how to use nuclear energy in safer ways.

When Jimmy Carter took office, the energy crisis was in full swing as trouble in the Middle East—including an Arab oil embargo (in retaliation for U.S. support for Israel) and then the Iranian Revolution—caused gasoline to be in short supply.

In 1977, he declared that the energy crisis was the "moral equivalent of war" and would require a similar sort of public mobilization. Recalls Jefferson Decker of *The Nation* magazine, "Carter asked

Americans to make a variety of personal sacrifices—'to take no unnecessary trips, to use carpools or public transportation whenever you can, to park your car one extra day per week, to obey the speed limit, to set your thermostat to save fuel'—in the hopes that the country could conserve its energy independence. Carter believed that, over the long haul, higher fuel prices would also be necessary to spur efforts to improve fuel efficiency and to invest in alternative technologies. [But] Carter faced a public that had come to believe that it had a right to cheap and plentiful gas—and expected the government to make it flow. Liberals in Congress hammered the President on the issue. Senator Ted Kennedy complained that Carter would 'force the poor to choose between food, medical care, and keeping warm' in the winter. Telling Americans that they were supposed to make do with less was not an option." (Kennedy would ultimately run against Carter in the Democratic primaries, and the 1980 Democratic National Convention was one of the nastiest on record. Although Carter was re-nominated, Kennedy had weakened him, giving rise to Ronald Reagan).

For the first time the price for a gallon of gas rose above a dollar, causing "panic at the pump." Service stations ran short of gasoline. Panicky drivers made the problem worse by repeatedly topping off their tanks in anticipation of future shortages, which clogged gas stations and forced other drivers to waste fuel waiting in line for hours. Violence broke out, and people even stole gas from each others' tanks.

In 1979 in Levittown, Pennsylvania about 100 long-haul truckers parked their rigs on Interstate 80 on the Pennsylvania side of the Delaware Water Gap and effectively shut down traffic on a well-traveled section of the transcontinental expressway. They were protesting high fuel prices and shortages and demanding relief. This evolved into an unofficial strike by tens of thousands of other truck drivers.

There were even riots, with protesters setting cars on fire and chanting, "More gas! More gas!"

At least one Democratic Senator, Henry "Scoop" Jackson, was visionary enough to define the problem, and warned about our nation's dependence on foreign oil. "We need to ask whether we must put ourselves in hock to Middle Eastern sheikdoms to keep roads clogged with gas-hungry automobiles," Jackson said.

But Meg Jacobs, who teaches history at Columbia and Princeton, says that "Most American consumers, who had become used to living with abundance, showed no interest in changing their patterns of consumption. They wanted their cars, they wanted their homes, and they wanted to continue living a life that would require lots of oil."

President Carter tried to address this issue head-on, when he made his so-called "crisis of confidence" speech in a televised address from the White House in July 1979. "It's clear that the true problems of our nation are much deeper—deeper than gasoline lines or energy shortages, deeper even than inflation or recession," he told viewers. "In a nation that was proud of hard work, strong families, close-knit communities, and our faith in God, too many of us now tend to worship self-indulgence and consumption."

His speech went over like a lead balloon. Even powerful Democrats, such as Tip O'Neill, showed no interest in promoting energy conservation. Instead, they concentrated on trying to pass measures in Congress that would help Americans pay their high fuel bills. This was one of the worst economic challenges since the Great Depression.

The public was "baffled," former Nixon speechwriter David Gergen reflected. They wondered: "Why should the most prosperous people in the world face [the] prospect of freezing homes and empty gas pumps?"

One bold stroke that Carter

President Jimmy Carter

took was to tax the "windfall profits" of the energy companies that benefitted whenever rates went up. The oil industry was also clearly engaged in price-fixing and other unfair business activities. Carter also signed the Emergency Natural Gas Act and created a Cabinet-level Department of Energy.

President Carter embraced renewable energy long before most of the country. He even had 32 solar panels installed atop the White House in 1979.

He once philosophized, "Like music and art, love of nature is a common language that can transcend political or cultural boundaries."

He believed that God gave Man dominion to care for His nature. He explained, "I was born into a Christian family, nurtured as a

Southern Baptist, and have been in weekly Bible lessons all my life.

"At least one Sunday each year was devoted to protection of the environment, or stewardship of the earth. My father and the other farmers in the congregation would pay close attention to the pastors' sermons, based on such texts as 'The earth is the Lord's, and the fullness thereof.' When humans were given domination over the land, water, fish, animals, and all of nature, the emphasis was on careful management and enhancement, not waste or degradation."

As President, Jimmy Carter increased the budget for the EPA and addressed specific environmental issues through his administration. Most notable was his signing of a bill that created a Superfund, which established a system of insurance premiums collected from the chemical industry to clean up toxic wastes. Carter predicted "This new program may prove to be as far-reaching and important as any accomplishment of my administration. The reduction of the threat to America's health and safety from thousands of toxic-waste sites will continue to be an urgent but bitterly fought issue—another example for the conflict between the public welfare and the profits of a few private despoilers of our nation's environment."

An avid fly fisherman who even tied his own flies, Carter wrote a book on the subject of enjoying nature, called *An Outdoor Journal: Adventures and Reflections.*

Robert Strong, Professor of Politics at Washington and Lee University praises the fact that in his post-Presidential years, Jimmy Carter "has emerged as a champion of human rights and worked for several charitable causes. To that end, Carter founded the Carter Presidential Center at Emory University in Atlanta, Georgia. The center, began in 1982, is devoted to issues relating to democracy and human rights. Additionally, Carter worked with Habitat for Humanity International, an organization that works worldwide to provide housing for underprivileged people." Many people across the political spectrum have called Jimmy Carter our greatest living former President.

4.

Reagan Was the Turning Point; A GOP War on the Environment as Democrats Fight Back

T HE ADMINISTRATION OF RONALD REAGAN USHERED IN A SERIES of Republican Presidents whose policies have indisputably created or at least exacerbated many of the environmental problems we face today. Reagan's own record will be remembered as one of the worst of any modern Presidency. This was a man who once claimed trees cause pollution. His goal was to dismantle environmental (and virtually all government) regulation and give away public lands and resources to big corporations and wealthy private entities. His adminstration left many problems unaddressed.

A former Grade B movie actor with a publicly sunny disposition and rouged cheeks, who called his wife, Nancy, "Mommie," Reagan was a bobble-headed "amiable dunce," as the late British Prime Minister Margaret Thatcher once described him. A rigid, rightwing doctrinaire, he believed what he believed—the facts be damned—and there was no changing his mind. Period.

Natalie Goldstein wrote in *Global Warming*: "Reagan's political philosophy viewed the free market as the best arbiter of what was good for the country. Corporate self-interest, he felt, would steer the country in the right direction."

He demonized government time and again, telling Americans that "Government is not the solution to our problem, government *is* the problem." This was a diabolical strategy to rationalize the lifting of important regulations that protected our citizens, so corporations could pursue capitalism without oversight or restraint. He privatized a lot of government functions, paying contractors to do work that federal agencies had formerly done. The result was not an improvement.

As Jeff Sovern, a professor of law at St. John's University School of Law observes, "Capitalism may be the best economic system ever devised, but one of its drawbacks is that it provides financial incentives to harm and even kill people. Just ask those people who say they have been victimized by cigarettes, predatory lenders, Volkswagen diesel emissions, Takata airbags, General Motors ignition switches, Trump University, Vioxx, asbestos or other products.

"With some of those examples, providers knew of problems long before they were disclosed but kept selling their wares, sometimes even covering up problems, all for profit.

"The more we discard regulation," Sovern concludes, "the more consumers must depend on companies to protect us from risks from their products that consumers cannot readily understand or don't have time to study. And as the [previously mentioned] examples indicate, companies sometimes succumb to the incentive to dispense with that protection."

Ronald Reagan popularized greed. He stoked public resentments over government programs that help people, citing "welfare queens" with his apocryphal and racist dog-whistle stories about black women living on the public dole but driving Cadillacs. He once tried to classify ketchup as a vegetable on the school lunch menu as he cut

When Governor of California, Reagan gave the middle finger to peaceful protestors. In 1983 he made this obscene hand gesture toward press photographers.

funding for meals even while the number of children living in poverty nearly doubled during his eight years in office. He suggested that one cause of the decline in public education was the schools' efforts to comply with court-ordered desegregation. (Reagan had opposed the Civil Rights Act of 1964 and the Voting Rights Act of 1965 signed into law by President Lyndon Johnson.)

Ronald Reagan made cuts to Social Security, Medicaid, Food Stamps, and federal education programs. He provoked bipartisan outrage when he attempted to purge tens of thousands of disabled people from the Social Security disability rolls, whom the Administration alleged—without proof—were not truly disabled. He slashed the budget for the Environmental Protection Agency by 22 percent.

Reagan apologists like to brag that he brought prosperity to America, but the combination of significant tax cuts (which a Democratic Senate helped enable) and a massive increase in Cold War related defense spending resulted in large budget deficits, an expansion in the U.S. trade deficit, as well as the stock market crash of 1987, while also contributing to the Savings and Loan crisis. The ultimate cost of the S&L debacle totaled around $150 billion, about $125 billion of which was directly subsidized by the taxpayers. Respected economist John Kenneth Galbraith called it "the largest and costliest venture in public misfeasance, malfeasance and larceny of all time."

None of this should be surprising. Reagan was one of the laziest of our Presidents up to that time. He told his aides that, rather than read his briefing books, he spent the eve of an economic summit watching *The Sound of Music*. "I put them aside and spent the evening with Julie Andrews," he confessed, without a trace of guilt.

In order to cover his federal budget deficits, the United States borrowed heavily both domestically and abroad, raising the national debt from $997 billion to $2.85 trillion during his eight years in office, and thus Reagan transformed the U.S. from being the world's largest international creditor to the world's largest debtor nation.

Reagan prevented his Surgeon General, C. Everett Koop, from speaking out about the AIDS epidemic. But Koop did an end-run around the President and in 1988 took the unprecedented step of mailing AIDS information to every U.S. household. This included recommending the use of condoms to help prevent infection with the disease.

The New York Times reported "With his 1981 appointments of two aggressive champions of industry, James Watt as Secretary of the Interior and Anne Burford as Administrator of the Environmental Protection Agency, Mr. Reagan seemed to have selected the nation's envi-

Interior Secretary James Watt was later charged with 24 felonies.

ronmental policies as a prime target of his social revolution. In its early years, the Administration moved rapidly to slash budgets, reduce environmental enforcement and open public lands for mining, drilling, grazing and other private uses."

Reagan was lax in enforcing antipollution laws and reckless in making public lands and resources available to profit-making corporations. "Environmental and conservation agencies were starved for money, the agencies were politicized and their staffs were demoralized...

"Worst of all, the administration deliberately delayed attacking long-term problems like global warming linked to pollution, acid rain, toxic waste, air pollution and the contamination of underground water supplies," the *Times* concludes.

Secretary of Interior James Watt tried to sell large quantities of Federal coal at prices well below market levels but was stymied. The Administration's plan to lease the entire Outer Continental Shelf for oil drilling ran into stiff opposition from members of Congress who said it could prove an environmental disaster. Watt wanted 80 million acres of undeveloped land in the United States to all be opened for drilling and mining by 2000. The area leased to coal mining quintupled during his term as Secretary of the Interior. Watt boasted that he leased "a billion acres" of coastal waters, even though only a small portion of that area would ever be drilled. Watt once stated, "We will mine more, drill more, cut more timber," and justified it on the basis of his being a "Dispensational Christian" and that we don't know when the Second Coming of Christ will be so we may as well exploit our resources now.

Sec. James Watt asserted that "the Department of the Interior ...must be...the Amicus for the minerals industry...in Federal policy." He commented "If you want an example of the failure of socialism, don't go to Russia, come to America and go to the Indian reservations." Watt mocked affirmative action with his description of a department coal leasing panel: "I have a black, a woman, two Jews and a cripple. And we have talent." He banned the musical group The Beach Boys from performing a Fourth of July concert on the National Mall, claiming they attracted "the wrong element" and encouraged drug use. He booked Reagan supporter Wayne Newton instead, who was booed when he took the stage.

Mr. Reagan had hired Ann Burford not to run the Environmental Protection Agency but to gut it. (Burford is the mother of current Associate Justice of the Supreme Court Neil Gorsuch whom Donald Trump nominated). She hired staff from the industries they were supposed to be regulating. In 1982, Congress charged that the EPA had mishandled the $1.6 billion toxic waste Superfund and demanded records from Burford. She refused and became the first agency director in U.S. history to be cited for contempt of Congress.

It was later discovered that the Administration was misusing Superfund grants intended for cleaning up local toxic waste sites to instead enhance the election prospects of local officials aligned with the Republican Party.

Reagan hired Donald Hodel as Energy Secretary, whose biggest claim to fame is that he idiotically suggested that the way to deal with ozone depletion was to apply stronger suntan lotion.

President Reagan dismissed proposals to halt acid rain, finding them burdensome to industry and refused to allow the EPA to try to reduce acid rain. He also questioned scientific evidence on the causes of acid rain.

In 1986, Reagan removed the solar panels that his predecessor Jimmy Carter had installed on the roof of the White House's West Wing, scoffing at their usefulness and calling them "just a joke."

Ultimately, Interior Secretary James Watt was embroiled in a controversy involving the department of Housing and Urban Development (HUD) whereby administration staffers defrauded the U.S. government out of money intended for low income housing. It resulted in six convictions, including Mr. Watt, who was indicted on 24 felony counts and pleaded guilty to a single misdemeanor. He was sentenced to five years probation, and ordered to pay a $5,000 fine—the sort of slap-on-the-wrist sentence only a white rich person could get.

To Peter Berle, former president of the National Audubon Society, "the biggest lost opportunity" of the Reagan years was "a failure to create a national energy policy built on conservation and energy efficiency. Instead, the Administration sought to rush the transfer of coal and oil from public lands and waters into private hands at bargain prices."

"It was eight lost years—years of lost time that cannot be made up and where a lot of damage was done that may not be reparable," said George Frampton Jr., former president of the Wilderness Society.

But former Gov. Bruce Babbitt of Arizona, a forceful advocate of environmental protection who ran unsuccessfully for the Democratic Presidential nomination, saw the glass as half full, not half empty. He said one of the "ironies of the Reagan administration's handling of the environment was the energizing of the environmental community and the rapid growth of a grass-roots environmental movement."

George H.W. Bush wasn't much better than Reagan

In 1988, Republican George Herbert Walker Bush stood on the deck of a boat in Boston's filthy harbor and proclaimed himself an environmentalist. Throughout the campaign he promised to scrub the skies with a strengthened Clean Air Act, set a goal of "no net loss of wetlands" and do everything in his power to halt global environmental threats. But he reneged on his promises when he got elected.

This was not surprising for a man who beat the superior Democratic candidate for President, Massachusetts Gov. Mike Dukakis by hiring a truly loathsome GOP dirty trickster to run his 1988 campaign, Lee Atwater, whose monstrously evil actions on behalf of other Republicans, including Ronald Reagan, were already well known. With the help of racist TV commercials like the infamous "Willie Horton ad," Bush overcame a 17 point lead by Dukakis to win.

His Administration proposed to open millions of acres of wetlands to development and to officially lower the risk estimates for dioxin and other long-feared contaminants. While Bush bragged about a new Clean Air Act that required companies to get permits that set limits on the levels of pollutants they were allowed to emit, his White House wrote regulations that would allow companies to exceed those

George H.W. Bush chose GOP dirty trickster Lee Atwater (right) to run his 1988 Presidential campaign. At the time, Atwater was a partner in a lobbying firm with Paul Manafort and Roger Stone (left and center), who have since been charged with felonies related to Donald Trump and Russian interference in the 2016 election.

limits automatically if states did not review requests for changes within seven days. Thus Bush created one giant loophole for his corporate friends while publicly pretending to care.

In the national energy strategy made public in the Persian Gulf War, George H.W. Bush called for increased production by drilling for oil in the previously untouched Arctic National Wildlife Refuge in Alaska while at the same time asking for very modest energy conservation measures.

In the debate over global climate change, Bush put the United States at odds with most of its allies, by refusing to impose limits on the production of carbon dioxide, the principal cause of global warming, because of concerns about curtailing industrial expansion.

After coming under heavy political pressure from developers, oil companies and other industries that contribute heavily to the Republican Party, the President proposed action that could reduce by tens of millions of acres the amount of low-lying bottomlands, marshes, swamps and other wetlands that had been protected from development. The Environmental Defense Fund said the proposal "would constitute probably the largest weakening of an environmental regulation in U.S. history."

This also was not surprising because George H.W. Bush had

been an oil man. In 1953, he borrowed today's equivalent of $8 million—half of it from his uncle Herbie—to start Zapata Petroleum Corp. and lease land in Texas to drill 71 wells in the hope of finding oil, then struck it rich. Not a single one had come up dry. There is also strong evidence to suggest that at the time, George was working with the Central Intelligence Agency (CIA)—though he denied it—despite the fact that later in his career, he was even named the director of that agency. So his affinity for protecting oil interests, and his proclivity to be duplicitous were both well established traits in this man—despite the lionization of him that has occurred in recent years by those who portray him as a WWII hero but ignore other aspects of his personality.

President Bush was also heavily influenced by his chief of staff, John Sununu, who—in November 1989—prevented the signing of a 67-nation commitment to freeze carbon dioxide emissions, with a reduction of 20 percent by 2005. Sununu is also the guy who launched the diabolical disinformation campaign that continues to this day, to confuse the public on the subject of global warming, by changing it from an urgent, nonpartisan and unimpeachable issue to a political one.

Sununu—who became well-known for his caustic attacks on Democrats—reportedly took personal trips, for skiing and other purposes, and classified them as official, for purposes such as conservation or promoting the Thousand Points of Light. An investigative report

John Sununu and President George H.W. Bush in the Oval Office.

by *The Washington Post* stated that government planes "took [Sununu] to fat-cat Republican fund-raisers, ski lodges, golf resorts and even his dentist in Boston." Sununu had paid the government only $892 for his more than $615,000 worth of military jet travel. After being forced to reimburse the Treasury Dept., he paid only about ten percent of what it actually cost for government transit to shuttle him around.

Bush's Vice President, Dan Quayle—whose embarrassing lack of intellectual prowess was evinced in repeated public gaffes—also became an influential player in environmental issues as head of the White House Council on Competitiveness, which emerged as an important voice for business interests.

If there was a hero in the Bush administration, it was EPA Administrator William Reilly who, in 1989, led the successful efforts to block the construction in Colorado of the Two Forks Dam, which would have destroyed a scenic canyon. Reilly was praised for helping to reach an agreement to reduce haze in the Grand Canyon. The EPA head also persuaded George Bush to authorize the Energy Department and the Pentagon to spend almost $7 billion for cleaning up toxic chemicals and radioactive contamination from weapons manufacture, and to sign a treaty banning imports of ivory.

Overall, however, President Bush did far less than he could have, if he hadn't been so concerned about alienating the GOP's business constituency.

Bill Clinton provided enlightened leadership on the environment

Martin Nie writes that "As in all Presidential administrations, [Bill] Clinton did not inherit a *tabula rasa* regarding environmental policy. A number of variables either constrained or facilitated Clinton's environmental agenda. First, although he garnered only 43 percent of the popular vote in 1992, people's environmental expectations of Clinton were very high. Surviving 12 years of Republican Presidential dominance, environmentalists and the public-at-large had much more confidence in the Democrats concerning the environment than in the GOP. Sixty-four percent of the electorate had more confidence in the President and the Congressional Democrats compared to 18 percent in the Congressional Republicans to handle environmental issues."

The best thing about Bill Clinton's administration was his Vice President, Al Gore, who had been a champion of the environment, albeit an inconsistent one. During the campaign, the two made it a central theme. Responding to Bush's extensive use of environmental symbolism over substance, Clinton pronounced "the days of photo-op

The 42nd President, Bill Clinton worked to balance environmental protection with the needs of a prosperous American economy.

environmentalism are over."

Clinton asserted that "I have never believed we had to choose between either a clean and safe environment or a growing economy. Protecting the health and safety of all Americans doesn't have to come at the expense of our economy's bottom line. And creating thriving companies and new jobs doesn't have to come at the expense of the air we breathe, the water we drink, the food we eat, or the natural landscape in which we live. We can, and indeed must, have both."

This from a man who, as Governor of Arkansas, had courted the poultry industry—a world-class polluter of that state's rivers and streams; a man who, by his own admission, winked at pollution to attract industry and jobs to a poor state.

When Clinton took office in 1993, Bush's lethargic federal clean up efforts had left 88 percent of the worst 1,200 toxic waste sites and their communities polluted after 12 years of federal efforts. Nearly 40,000 urban industrial sites sat abandoned with no federal strategy to redevelop them. Sixty-two million people lived in areas with drinking water below federal standards, and nearly 157 million people—62 percent of the country—breathed air that failed to meet federal standards.

The Clinton administration strengthened the Safe Drinking Water Act, requiring America's 55,000 water utilities to provide regular reports to their customers on the quality of their drinking water.

Clinton adopted the toughest standards ever on soot and smog, ordered major reductions in tailpipe emissions from cars, light trucks and sport utility vehicles, and mandated reducing the level of sulfur in gasoline by 90 percent.

President Clinton initially promised to enforce environmental laws with jail terms for corporate polluters when necessary. Over 200 criminal environmental cases were pursued in fiscal year 1994. Criminal charges were brought against 250 individual and corporate defendants with 99 years in jail sentences and $36.8 million in criminal fines assessed that year. By the fall of 1995, however, the EPA was forced to raise the standards of cases that it would prosecute to "significant and egregious cases" because of scarce resources.

Still, in 1999 alone, the EPA assessed a total of $228.3 million in civil and criminal penalties, the most ever assessed and $87 million more than in 1992. The EPA referred 241 criminal cases to the Justice Department in 1999—more than twice the number referred in 1992. At least 322 defendants were charged in 1999, and 2,500 total months of sentences were handed down, more than doubling enforcement activity in each category over 1992 levels.

Throughout his Presidency, Bill Clinton stood firm against Republican attempts after the 1994 elections to roll back environmental laws and regulations through the appropriations process. He routinely vetoed budget bills saddled by GOP anti-environmental riders aimed at avoiding public scrutiny. Yet he took his own back door approach to environmental protections, using executive orders to create 17 new national monuments, and expand four more, without Congressional approval. These monuments preserve more than 4.6 million acres in the lower 48 states.

Clinton chose to focus largely on developing alternative energy sources, rather than new sources of fossil fuels. His administration launched more than 50 major initiatives to improve energy efficiency and develop clean, renewable energy sources. In just three years, the President secured more than $3 billion—a 50 percent increase in annual funding—to research and develop clean energy technologies.

In Clinton's eight years in office, three times as many Superfund sites were cleaned up as in the previous 12 years. Cleanup was completed or underway at 92 percent of all Superfund sites.

Even the federal government became more efficient, reducing its annual energy bill by $800 million in 1999 alone. The Clinton administration implemented new energy efficiency standards for heating and cooling equipment, water heaters, lighting, refrigerators, clothes washers and dryers, and cooking equipment, with most of these new regulations still in force today.

In the Pacific Northwest, after promising reform, Clinton's worst environmental blunder was reopening some forests to logging. Yet during his watch, wolves came back to Yellowstone, fishing was banned to save endangered salmon, and a Grand Canyon river was flushed clean.

Clinton warned that the new global economy requires the United States to take a leadership role in combating "the degradation of the global environment," as well as the spread of "deadly weapons and disease." But among his most controversial moves was the opening of permanent normal trade relations with China in May 2000. Opponents cited China's abysmal record on environmental issues, and warned that ending the practice of annually reviewing China's record before approving trade removes one of America's only means of influencing China's policies.

Bill Clinton also encountered criticism over the U.S. relationship with the World Trade Organization, an international coalition

Vice President Al Gore applauds after Bill Clinton signs a bill in September 1996 designating 1.7 million acres of land in southern Utah's red-rock cliff as the Grand Staircase-Escalante National Monument, at the Grand Canyon in Arizona.

that has been condemned for its lax policies toward environmental protection and human rights.

Despite the laudable achievements of the Clinton/Gore administration, "setbacks for the greens came at a dizzying pace," according to Alexander Cockburn and Jeffrey St. Clair in *Al Gore: A User's Manual.* They write "Tax breaks were doled out to oil companies drilling in the Gulf of Mexico. The Department of Agriculture okayed a plan to increase logging in Alaska's Tongass National Forest, the nation's largest temperate rainforest. The Interior Department, under orders from the White House, put the brakes on a proposal to outlaw the most grotesque form of strip mining, the aptly-named mountaintop-removal method. With Gore doing much of the lobbying, the administration pushed through Congress a bill that repealed the ban on the import of tuna caught with nets that also killed dolphins. The collapse was rapid enough to distress so centrist an environmental leader as the National Wildlife Federation's Jay Hair, who likened the experience of dealing with the Clinton-Gore administration to 'date rape.'"

But these issues, while attracting vocal and highly visible protests, are less likely to be remembered as President Clinton's pro-environment choices: the Grand Staircase-Escalante National Monument; acres of roadless, old growth forest; groves of wind turbines producing pollution-free electricity.

Ironically, the most pro-environment candidate for the Presidency in the election of 2000, consumer advocate and Green Party nominee Ralph Nader, ended up being the "spoiler" who cost Al Gore the White House. Nader won about three percent of the nationwide vote, including more than 97,000 votes in Florida—the state on which the election ultimately turned—thus bequeathing us "Dubya" Bush, son of the first President Bush.

George W. Bush was an unmitigated disaster

The tone was set in the first 100 days when George W. Bush abandoned a campaign promise to regulate carbon dioxide from coal-burning power plants, the biggest contributors to global warming. Days later, the White House announced that America would not implement the Kyoto global climate change treaty—that had been signed by 37 industrialized countries and the European Community committed to reduce greenhouse gas emissions

The two moves made clear that this President—a former oil man like his father—cared more about protecting America's coal and oil industries than he did the good of the country.

President George W. Bush and Vice President Dick Cheney.

Christine Todd Whitman, who was the head of the Environmental Protection Agency at the time, later described the exit of Kyoto as "the equivalent to 'flipping the bird,' frankly, to the rest of the world."

But it was the manner of Bush's exit from Kyoto that provided the most sustained damage, say environmentalists, with the administration casting doubt on the science that demonstrated convincingly that global warming was a real problem.

"The idea of a head of state putting the science question on the table in the way that he did was horrifying to most of the rest of the world," said Eileen Claussen, president of the Pew Center on Global Climate Change.

The Guardian newspaper in the UK said that "The disinformation campaign became a defining element of the Bush era—and was perhaps the most damaging."

"Certainly the most destructive part of the Bush environmental legacy is not only his failure to act on global climate change, but his administration's covert attempt to silence the science alerting us to the urgency of the problem," said Jonathan Dorn of the Earth Policy Institute in Washington.

"The campaign to keep the public unaware of the evidence on climate change came to light in October 2004 when the NASA scientist, James Hansen, accused the Bush administration of trying to block data showing an acceleration in global warming," *The Guardian* recounts.

"The full extent of the White House efforts to downplay, distort and outright censor the science on climate change remains unclear—but such efforts continued even after Hansen accused the Bush administration of censorship."

In July 2008, Jason Burnett, a former official at the EPA, wrote a letter to the United States Senate revealing efforts by the office of Vice President Dick Cheney, and the White House Council on Environmental Quality to censor discussion of the consequences of climate change.

Burnett said the White House tried to circumvent a 2007 Supreme Court decision compelling the EPA to regulate car emissions by doctoring scientific findings on the costs of fuel-efficiency standards. The White House objected to a study showing that the benefits of raising fuel standards outweighed the costs.

Meanwhile, Bush officials worked aggressively to undo decades of important regulations protecting our environment. One big target was the Endangered Species Act, because they feared it could be used to force limits on greenhouse gas emissions.

An official report by the U.S. Inspector General found widespread political interference in the management of endangered species. The report said that the Deputy Secretary of the Interior, Julie MacDonald—who headed the endangered species program at the U.S. Fish and Wildlife Service—intervened improperly in 13 of the 20 cases under investigation, overruling the recommendations of field biologists that species be protected.

In 2008, Dick Cheney sought to doctor testimony prepared for a Senate hearing on California's efforts to impose stricter fuel efficiency requirements than the national standard.

Former First Lady Barbara Bush, mother of George W., freely admitted that she had repeatedly expressed concerns to her son about the negative influence Dick Cheney had over him, and that the Vice President was pulling Dubya too far to the right on some issues.

Other nefarious actions by the Bush White House included:

• Gutting key sections of the Clean Water and Clean Air acts.

• Dismantling the protections of the Endangered Species Act.

• Opening millions of acres of wilderness to mining, oil and gas drilling, and logging.

• Defunding programs charged with the clean-up of toxic industrial wastes such as arsenic, lead and mercury.

• Reducing the enforcement effort in the Environmental Protection Agency.

• Removing grizzly bears and wolves from the endangered species list.

• Endorsing commercial whaling.

• Approving mountain-top removal for coal mining.

Bush even issued an 11th hour flurry of orders allowing oil companies to be able to drill within sight of the Arches National Park in Utah, excusing Federal agencies from having to consult with government wildlife experts when they open up new areas for logging or road construction, and barring the EPA from looking at the effects of global warming on protected species.

But the overwhelming environmental and human catastrophe to occur during the Bush years was the terrorist attack on September 11, 2001. *Politico* reported that "months earlier, starting in the spring of 2001, the CIA repeatedly and urgently began to warn the White House that an attack was coming..." But Vice President Dick Cheney questioned the accuracy of these reports. After CIA analysts did more research, they produced a report titled "UBL [Usama Bin Laden] Threats Are Real." President Bush's response was to tell the CIA, "All right. You've covered your ass." He did nothing.

Politico quotes former CIA director George Tenet as saying "And the word back...'we're not quite ready to consider this. We don't want the clock to start ticking.' Translation: they did not want a paper trail to show that they'd been warned."

On July 10, 2001, CIA officials urgently went to see Bush's National Security Adviser Condoleezza Rice and once again emphasized the imminent terrorist attack. But she played down the threat of stateless terrorism and later claimed she did not even recall the specific meeting. When subpoenaed to testify to Congress, she refused. *Politico* quoted Cofer Black, then chief of the CIA's counter-terrorism center, as saying "To me it remains incomprehensible still. I mean, how is it that you could warn senior people so many times and nothing actually happened? It's kind of like *The Twilight Zone*."

The President's Daily Brief (PDB) prepared by the CIA on August 6, 2001 and given to George Bush at his Texas ranch while he was on vacation carried the warning: "Bin Laden Determined To Strike in U.S." It outlined "patterns of suspicious activity in this country consistent with preparations for a hijacking of U.S. aircraft." Rice later claimed the warning contained old information, but Sean Wilentz of *Salon* magazine suggested that the PDB contained current information based on continuing investigations.

Ironically, the North American Aerospace Defense Command (NORAD) conducted a training exercise two years prior to 9/11 in which it simulated a civilian airliner being hijacked and crashed into buildings, including the World Trade Center. In October 2000 the

The worst attack on America in history occurred on the watch of Bush and Cheney, who failed to heed advance warnings and to keep us safe.

U.S. Defense Dept. had even conducted exercises rehearsing a plane crashing into the Pentagon. And the National Reconnaissance Office had scheduled an exercise for Sept. 11, 2001 simulating a crash of a jet into one of its buildings.

Then all of these things happened for real. Why weren't we prepared?

When the twin towers of the World Trade Center in New York City were attacked on September 11, 2001 by hijacked planes crashing into them—plus two other airliners, including one that hit the Pentagon in Washington—more than 3,000 people were killed (including at least 400 police officers and firefighters). It was the deadliest terrorist act in U.S. history and the most devastating foreign attack on American soil since Pearl Harbor.

While there are thousands of known victims and survivors, some remain unknown. The Office of the Chief Medical Examiner of the City of New York has custody of 7,930 unidentified remains of those killed in the attacks. The remains are located in the World Trade Center Repository situated between the two footprints of the Twin Towers on the sacred ground of the World Trade Center site.

The collapse of the 110-story complex created huge dust clouds that rolled through the streets of Lower Manhattan, breaking windows and forcing dust and debris into the interior of surrounding buildings. Chemicals and materials present in the billowing clouds included pulverized plaster, paint, foam, glass fibers and fragments, fiberglass, cement, vermiculite (used as a fire retardant instead of asbestos), chrysotile asbestos, polycyclic aromatic hydrocarbons (PAHs), polychlorinated biphenyls, polychlorinated dibenzodioxins, polychlorinated dibenzofurans, pesticides, phthalate esters, brominated diphenyl ethers, cotton fibers and lint, tarry and charred wood, soot, rubber, paper and plastic.

Combustible materials buried in the rubble of the towers provided fuel for a fire that burned until December. Many of the materials were made of substances that when burned release highly toxic fumes. According to Thomas Cahill, a professor of physics and engineering, "The debris pile acted like a chemical factory. It cooked together the components and the buildings and their contents, including enormous numbers of computers, and gave off gases of toxic metals, acids, and organics for at least six weeks."

Computers and fluorescent lights contained mercury. There was a Secret Service shooting range that kept millions of rounds of lead

People covered in toxic soot fleeing the scene of the World Trade Center attacks.

ammunition on hand. And a U.S. Customs lab had in its inventory thousands of pounds of arsenic, lead, mercury, chromium, and other toxic substances. Other products in the buildings included synthetic fabrics, plastics, laminates, the di-electric fluids that encase electrical cables, capacitors, electrical cable insulation and transformers. The toxins resulting from the combustion of these materials include toxic lead, volatile organic compounds, dioxins and more.

The plume from the buildings' collapse moved from the WTC site in Lower Manhattan directly over Brooklyn suburbs.

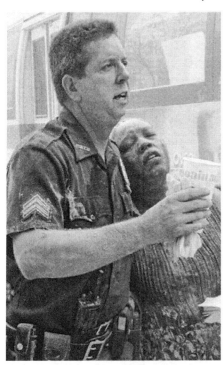

Sgt. Marty Duane was one of the many first responders who wore no protection from the dangerous air.

George W. Bush was credited with going to the scene, standing on a pile of rubble, and using a megaphone to promise the terrorists would be punished. But they weren't; he launched a disastrous pre-emptive attack on Libya's Saddam Hussein instead, because, George said, "he tried to kill my daddy," even though Saddam had nothing to do with the attacks. It was the Saudis who attacked our nation, and all Bush did is de-stabilize the Middle East to this day.

The Bush administration also failed to adequately protect the first responders at the scene of the New York City attack and one at the Pentagon in Washington. The blame for the NYC problems must also be shared by then-Mayor Rudolph Giuliani.

During the first 25 days of the rescue/recovery effort at the World Trade Center site, 800 policemen were provided with only paper masks. Printed on each of the masks was a disclaimer stating: "Warning, this mask does not protect your lungs." Almost 4,000 firefighters who participated in the rescue efforts at the World Trade Center complained of problems including trouble breathing, wheezing, coughing, severe eye irritation, headaches and stomach problems.

Nicole Pollier of the Center for Constitutional Rights, testified before the Environment Committee of the New York City Council at the time and said that "virtually none of the people working at the WTC disaster site are or have been wearing any personal protective equipment." Only five to ten percent of the workers wore disposable dust masks, she said. Additionally, "workers leaving the site are not decontaminated, nor do they use the washing stations that have been set up at the perimeter of the site by volunteer organizations." Pollier said that under Bush, OSHA took the position that the site's designation as a "search and rescue" operation denied it the authority to enforce safety laws.

The firefighters union claimed the illnesses could have been prevented if proper respirators had been provided to firefighters working at the World Trade Center site.

Over 1,700 police officers and firefighters filed lawsuits claiming that conditions at Ground Zero or the Fresh Kills landfill caused their illnesses, including sarcoidosis, asthma, reactive airway disorders, and chronic coughs. Cancer rates have tripled among First Responders. Neurological damage is also showing up.

The 2001 collapse of the twin towers produced a noxious fog of burning plastic, asbestos, fiberglass and jet fuel. This foul, carcinogenic mixture was inhaled not just by 9/11 responders, but by up to half a million people who lived in, worked in, attended school in or visited lower Manhattan in the ten months following September 11, 2001. Thousands of people have become ill with ailments like cancer that are thought to be attributable to the terrorist attacks. A victims compensation fund has run out of money, and claims keep pouring in even in 2019.

The terrorist attacks were not the only instances where the Bush administration had an inadequate response. Along came Hurricane Katrina. It was one of the worst natural disasters in our nation's history.

The hurricane—a Category Five storm—hit the U.S. Gulf Coast on the morning of Monday, August 29, 2005, spreading across 400 miles with sustained winds of up to 125 mph. A storm surge as high as 29 feet in some places rolled across levees and drainage canals and led to widespread flooding and the displacement of hundreds of thousands of people from their homes in Louisiana, Mississippi and Alabama. Damage was estimated at $100 billion and at least 1,836 people were killed from the hurricane or subsequent flooding.

As strange as it may seem in this era of instant communication and the 24-hour news cycle, President Bush didn't initially pay attention to the biggest news story of the moment because he was once again on vacation and allowed himself to get isolated from the country.

As *U.S. News & World Report* noted, "The National Weather Service had warned on August 28, the day before the storm made landfall on the Gulf Coast, that 'most of the [Gulf Coast] area will be uninhabitable for weeks...perhaps longer.' New Orleans Mayor Ray Nagin ordered the city evacuated and he opened the Superdome as a shelter, but the thousands of people who sought refuge there found little or no food, water and medical care.

"Americans across the country were shocked by the television images they saw in Katrina's immediate aftermath. People stood on rooftops waving their arms and pleading for help as the flood waters inundated their communities. Desperate folks in the Superdome appeared in heartbreaking TV interviews begging for aid in their time of need. Making matters worse was that 67 percent of New Orleans was African American and 30 percent of the residents were poor, creating the impression that the government was insensitive and neglectful of minorities and the less fortunate."

U.S. News states that "While all this was going on, the President of the United States remained aloof from the disaster. Day after day, George W. Bush continued a long-planned vacation at his 1,600-acre Prairie Chapel Ranch in Crawford, Texas, and his staff didn't want to

Coast Guard Petty Officer Shawn Beaty looks for survivors in the wake of Hurricane Katrina. Beaty is a member of a helicopter rescue crew sent from Florida to assist.

burden him with detailed information about the situation on the Gulf Coast. When Katrina made landfall, Bush had been on holiday at his ranch for 27 days, according to a tabulation kept by CBS News."

The Federal Emergency Management Agency (FEMA) was slow to respond and largely bungled the job. The Republican anti-government mantra—initially articulated by Ronald Reagan years earlier—was proven wrong in the most tragic way. Bush finally ended his vacation but as he headed back to Washington, merely flew over the ravaged area. Showing just out of touch he was, Bush then publicly praised FEMA director Michael Brown early in the crisis. "Brownie," Bush said, "you're doing a heck of a job."

Besides harming more than one million people, and killing over 250,000 pets, Katrina also had a profound impact on the environment. The storm surge caused substantial beach erosion, in some cases completely devastating coastal areas. The hurricane caused massive tree loss, oil spills, the closure of 16 National Wildlife areas and extensive other damage.

Twelve years after Katrina, numerous New Orleans neighborhoods suffered Katrina-level floods—including a few neighborhoods that had

Nero "fiddled while Rome burned," and George W. vacationed while Katrina victims drowned.

never flooded before—from a basic rainstorm. The culprit this time was faulty sewer drainage equipment. This means that, despite $15 billion worth of levee walls built around certain parts of New Orleans, the families in the city remain vulnerable to flood displacement. Rising sea levels caused by global warming only exacerbate the problem.

A summation of the George W. Bush years was provided by Josh Dorner, a spokesman for the Sierra Club, one of America's largest environmental groups. He said of Bush "He has undone decades if not a century of progress on the environment. [He] has introduced this pervasive rot into the federal government which has undermined the rule of law, undermined science, undermined basic competence and rendered government agencies unable to do their most basic function even if they wanted to."

The actions taken during "Dubya" Bush's eight years in the White House were seen by environmentalists as ideological rather than scientifically based. (They will be further detailed in later chapters).

But Bush's bad record served as a motivating force for change. As *The Guardian* chronicles, "The Bush administration's refusal to cap carbon dioxide emissions acted as a catalyst, with 24 states acting on their own to put in place regional cap and trade networks. Some 27 states enacted renewable portfolios, mandating local power companies to produce more of their electricity from sun, wind and solar power."

Perhaps as a sort of cleansing of his soul and a sense of guilt over how he had degraded the environment, just two weeks before leaving office, Bush designated nearly 200,000-square miles of the Pacific Ocean as national monuments. It was, said Joshua Reichert, managing director of the Pew Environment Group, "a significant conservation event."

Obama promoted clean energy alternatives

Barack Obama certainly achieved some laudable accomplishments with regard to the environment during his two terms as President, but he also missed some important opportunities.

Still, as environmental advocate Rick Smith observes, today's Green New Deal proposed investments "mirror many of the same projects funded by the Obama 2009 stimulus. More than $20 billion funded energy efficiency and conservation programs under Obama. Another $25 billion promoted renewable electricity generation. About $10 billion was dedicated to smart electric meter and smart grid technology as well as long-distance transmission lines. Advanced vehicle technologies such as electric battery manufacturing subsidies and heavy-duty diesel retrofits received over $5 billion. The Obama clean energy stimulus invested in $2 billion in state of the art carbon capture and storage technologies for coal-fired power plants."

Though anti-environment Republicans have mocked the Obama administration for financing a solar manufacturer called Solyndra that failed, the record is much more favorable than that.

"The results of Obama's spending produced some remarkable results," Smith asserts. He points to the fact that "America's wind capacity has more than tripled since 2009. Solar capacity is up more than [30 times]. LEDs have grown from one percent of the market in 2009 to more than 50 percent today. Plug-in electric vehicles were nearly non-existent in 2009 and now they number more than one million. Iowa is the poster child for the growth of wind energy with 37 percent of electrical generation coming from the wind.

"Republicans are crying wolf when they claim Americans can't achieve the goals of the Green New Deal. Obama proved we can, even as Republicans did everything in their power to block his clean energy initiatives.

"Just as Republicans savaged Obama's green energy investments, they are viciously attacking the Green New Deal... Obama's commitment to green energy set the stage for many of the Green New Deal proposals," Rick Smith concludes.

Douglas Brinkley, an author and historian at Rice University asserts that "It's a gargantuan legacy. I put [Obama] as one of the top environmental Presidents in history."

It's useful to review the historical context of Barack

Obama successfully promoted wind power.

Obama's two terms as President and how that influenced his ability to get green initiatives passed, and what other policy priorities distracted him from his eco-friendly impulses.

Barack Obama assumed power at a time when the U.S. economy was in free fall in the worst recession since the Great Depression of the 1930s. The economy was losing 800,000 jobs per month. Bush had squandered a $127.3 billion surplus that Bill Clinton had left, and turned it into a $269 billion budget deficit by giving massive tax cuts to the rich and launching a costly pre-emptive and illegal war in Iraq.

So as the incoming President, Mr. Obama was naturally concerned about righting the ship of state.

In his first two years, he was lucky enough to have both houses of Congress controlled by members of his own party, but to overcome then-Senate Minority Leader Mitch McConnell's filibusters, a 60-vote majority was required in the Senate, and that meant Obama needed to convince three Republican Senators to go along with his proposals.

That proved to be difficult, if not impossible in some instances. For example, Obama's economic stimulus proposal included $10 billion for a nationwide effort to upgrade the energy efficiency of public schools. But GOP Senator Susan Collins of Maine—who is often portrayed as a moderate but votes the radical rightwing party line when it really counts—opposed this sensible idea. So the legislation ended up with nothing for green schools.

Barack Obama came into office with certain distinct advantages. As the first African-American President, his public approval rating soared to 80 percent. But instead of taking advantage of this golden opportunity to press for the progressive agenda he promised to pursue—including important initiatives regarding climate change—he sought to garner the cooperation of Republicans by offering them what he called "compromises," but which some members of his own party came to view as "sell-outs."

Was Obama really that naive? According to *Frontline*, there was a major anti-Obama strategy meeting the night of his inauguration. The meeting was in a Washington D.C. steakhouse, organized by high-level pollster and consultant Frank Luntz, and included Paul Ryan, Eric Cantor, and Kevin McCarthy, among others. Mitch McConnell—who quite possibly is the most venal leader Congress has ever seen—vowed that the number one priority for Republicans was to make Obama "a one-term President."

On a number of occasions, members of the GOP even voted against their own legislation if President Obama supported it.

As David Durham of the online forum *Quoro* writes, "A pattern was set that played out as an effort to make our President fail no matter what. Is it good for the country? Yes, but it would be an Obama success and we can't have that so the country will just have to suffer. So no jobs bills. While we're blocking jobs bills we'll blame Obama for a slow recovery. Brilliant!

"After President Obama was able to get through a stimulus bill with absolutely no GOP support, Republicans would get their pictures taken with stimulus checks, taking credit for the very thing they voted against."

A fair criticism of Barack Obama is that he did not take advantage of Democratic majorities in Congress to do things like close Gitmo,

The President holds twins belonging to a White House staff member. As a father himself of two daughters, Barack spoke often about creating a better future.

stop mass surveillance, or create whistleblower protections.

Instead, as one of his first actions, the new President signed a bill rewarding Wall Street's evil doers with $350 billion of no-strings-attached bailout money and enormous legal latitude. Wall Street had been given the money with the idea that banks would help struggling borrowers, curtail home foreclosures and increase lending to stimulate the economy, but that's not what happened. Many of the financial institutions that got helped simply turned around and gave lavish bonuses to senior management, made "opportunistic acquisitions" and mergers, and screwed over their workers.

What is little known is that in 2008, while Bush was still President, the Federal Reserve gave banks a $7.7 *trillion* secret loan, which in turn helped the bankers rake in $13 billion in undisclosed profits. All the while, the major banks were lying to their investors. That dwarfs the Treasury Department's better-known $700 billion Troubled Asset Relief Program (TARP) which Republicans sanctimoniously attacked. If the public had known of the earlier loan, it could have led to a push to reinstate the Glass-Steagall Act, the Roosevelt-era measure which for decades had prohibited banks from owning investment companies and vice versa, thereby limiting their size and vulnerability to such financial meltdowns.

To win support for his urgent stimulus package, Obama had

to limit the amount of money allocated to jobs creation, and ruled out direct government jobs programs modeled on the 1930s-era Works Progress Administration he had promised during the campaign. These concessions were necessary to win the votes of exactly three Republican Senators, but no Republicans supported him in the House.

Yet Obama often capitulated to Republicans when he didn't have to. Lance Selfa, author of *The Democrats: A Critical History*, observes that "Using only executive action, Obama could have unwound the Bush era bailouts for the bankers and pressured bankruptcy judges to reduce or wipe out mortgage holders' debt. At the very least, he could have refused to allow executives from the bankrupt insurance giant AIG to collect their multimillion-dollar bonuses from the taxpayers' dime. Yet he did none of these things. Billions of dollars appropriated to help homeowners avoid foreclosure remained unspent." Over the next several years, the Obama administration helped drive the Republicans' austerity agenda that further cut into working-class living standards. Time and time again he rolled over for the GOP, making some wonder if he was a Trojan Horse.

As Lance Selfa writes, "In 2010, the administration and Congress agreed to extend the Bush administration's mega-tax cuts for the rich for two more years—something Obama had promised again and again that he would never do. In 2012, when the tax cuts were set to expire, Obama agreed to allow most of them to remain.

"In 2011, Tea Party-addled Republicans in the House threatened to plunge the U.S. government into default to force through huge cuts in government spending. Instead of calling their bluff, Obama agreed to $1.5 trillion in federal spending cuts over the next decade... During the negotiations over the 2011 government shutdown, Obama offered the Republicans unprecedented cuts to Social Security and Medicare."

Glenn Greenwald complains "in many crucial areas, he has done more to subvert and weaken the left's political agenda than a GOP President could have dreamed of achieving."

What to make of Barack Obama's actions? He had famously said: "Generations from now, we will be able to look back and tell our children...this was the moment when the rise of the oceans began to slow and our planet began to heal." But paradoxically, he did less than he could have to make good on his promises—until the waning days of his administration.

Instead, he carried on with "business as usual": Opening public lands to coal mining and increasing off-shore drilling. The latter had disastrous consequences when the British Petroleum Oil spill off our Gulf Coast became a monumental ecological and humanitarian catastrophe (*which will be discussed in a later chapter*).

Obama's two biggest environmental initiatives never materialized. Cap-and-trade died in the Senate. His clean power plan is ensnarled in lawsuits. Obama used executive powers to bypass Congress in slapping restrictions on dirty power plants, but first courtroom challenges placed it in limbo, and then along came Donald Trump to undo much of what Obama did.

As Marianne Lavelle elaborates in *Vox*, "Had the White House pushed for a comprehensive national climate plan early, it could have given Obama's climate agenda legislative backing, making it much harder for his successor to undo. A cap-and-trade bill, Waxman-Markey, based heavily on a proposal by a coalition of industry and environmental groups, had squeaked through the House in 2009. But after its Republican backers in the Senate got cold feet, Obama rallied no support behind it and it fizzled there. Obama never developed his own legislation proposal to replace it."

Sonya Diehn says that "To his credit—and what will likely comprise his actual environmental legacy—were hard-won stronger fuel efficiency standards, and investments in renewable energy."

EPA data showed that during Obama's time in office, carbon emissions decreased nine percent, while the U.S. economy grew more than ten percent. By comparison, the European Environmental Agency recorded a 14.7 percent in carbon emissions drop over the same period across the 28 EU member states.

Obama also had success in pushing solar and wind energy.

Rob Sargent of Environment America points out that "The Obama administration actively supported the offshore wind energy industry, and ultimately oversaw the first offshore wind farm coming online in 2016. And they aggressively pushed energy savings by finalizing 50 efficiency standards—more than any previous administration—saving consumers $550 billion on their utility bills."

Sargent also notes that "Land-based wind costs have dropped 41 percent since 2008, and wind now powers over 17 million U.S. households. The cost of rooftop solar has declined by 54 percent, and as of the beginning of 2016, one million households had solar panels. Finally, utility-scale solar costs have been cut by 64 percent, powering two million homes."

In the final months of his adminstration, Obama issued executive orders to weave climate considerations into everything from urban planning to national security. Clean power was the centerpiece, but opponents called it a "war on coal," and 27 states sought to block it through the courts.

As *Climate Change News* explains, "Nationally and internationally, climate policies tend to focus on where coal, oil and gas are burned, not where they are extracted. Yet energy companies mine and drill like there is no tomorrow, their business plans incompatible with 'safe' limits on global warming."

While this was not something Obama systematically addressed, decisions like a coal mine moratorium on public lands showed some sympathy with the keep-it-in-the-ground argument. In 2015 he vetoed the Keystone XL oil pipeline, a rallying issue for campaigners, on climate grounds.

In recent years those groups complained that timelines for greenhouse gas reductions were too long. They accused the administration of underfunding agencies that oversee endangered species protections. And they went to court to challenge sales of federally owned coal with no regard for future pollution.

Matthew Brown reported in the Associated Press that "The Obama administration's scramble to finalize key environmental policies

Obama squandered time when Democrats controlled Congress but at the 11th hour took executive action to protect the environment, which Donald Trump can undo.

in its last days obscures the fact that many of those actions were in the works for years." So why did they wait until the last minute?

To be fair, President Obama's signature legislative priority was enactment of the Affordable Care Act (ACA), which gave health insurance coverage for the first time to more than 23 million Americans and saved many lives. As mentioned in an earlier chapter, provisions of the ACA have also helped tens of millions more, such as the prohibition against insurance companies rejecting coverage of people with pre-existing conditions. There is also the untold story of how the often-ridiculed (now House Speaker) Nancy Pelosi revived Obamacare after Democrats left it for dead, and through brilliant strategizing and a tough-mindedness that shamed her weaker colleagues, showed them all how grit and determination can win the day.

Jacqueline Patterson, climate lead at the National Association for the Advancement of Colored People (NAACP), welcomed Obama's attention to those communities—often black, Latino or indigenous—on the front lines of both climate impacts and health threats. "That awareness wasn't always incorporated into the way policy was made," she comments.

There were also some notable successes in the Obama years with regard to green-friendly initiatives that received less attention, but that will be elaborated on in successive chapters of this book.

These include his signing of the Omnibus Public Land Management Act of 2009, which resulted in the President protecting over two million acres of wilderness.

He established the America's Great Outdoors Initiative in 2010, to promote recreation through various programs.

Obama expanded the California Great Coastal National Monument along a vast stretch of the western coastline.

To encourage children to visit National Parks, President Obama bestowed free admission to fourth graders.

Mindful of the fact that bees and butterflies are disappearing, and this could have a devastating effect on our nation's crops (since these insects are crucial pollinators), Obama formed a task force to develop a strategy to help the situation.

Barack Obama was above average in his environmental stewardship but he regrettably missed some opportunities to do more.

5.
Donald Trump, the Russians and A Bad Environmental Policy

A S BAD AS THE REAGAN AND TWO BUSH ADMINISTRATIONS were for the environment, Donald Trump's actions are exponentially, alarmingly worse—not just for Americans, but for the world at large. His corruption on all fronts makes him the most dangerous President in U.S. history.

Under pressure from critics for not formulating an energy policy, in early 2016 the Trump campaign brought in two men who had ties to Russia. They were Carter Page, who had worked with Russian intelligence officials and George Papadopoulos, whose interactions with foreign governments and claims about Hillary Clinton's emails ultimately ignited the most explosive scandal in American history.

It appears that Trump's environmental policies have been shaped, at least in part, by his allegiance to Russia and its interests.

The redacted 448-page report prepared by Special Counsel Robert Mueller, who spent two years looking into Russia's interference in our 2016 elections, reveals that Carter Page drafted the outlines of Trump's first major policy speech on the environment, delivered in May 2016 in Bismark, N.D., one of the nation's hubs for fracking.

In that speech, Trump promised to "cancel the Paris climate agreement" and to roll back President Obama's signature climate policy. He called Obama's priority on climate science "phony."

According to the Mueller investigation, starting in January 2016, Carter Page told top Trump campaign officials that he had repeated contacts with Russians with close ties to the Kremlin. In fact, he had even lived in Moscow from 2004 to 2007, and developed relationships with Russian oil executives, especially from the Kremlin-connected company Gazprom. He aroused the scrutiny of U.S. national security investigators as far back as 2013 because of his interactions with a Russian intelligence operative.

For any normal Presidential campaign, that sort of résumé

would have raised alarm. Instead, by March 2016, Page was officially named as Donald Trump's foreign policy adviser on Russia and energy issues. In July 2016, Page went to Moscow at Russia's invitation—on a trip approved by the Trump campaign—and delivered a Putin-friendly speech. Shaun Walker, the *Guardian's* Russia correspondent, said Page's presentation was really weird: "He was talking about the United States' attempts to spread democracy, and how disgraceful they were."

Before his connection to the Trump campaign, Page had been recruited by Russian intelligence as a foreign source, the Special Prosecutor's report states. He had experience on energy issues in Russia and Russians even gave Page false encouragement to believe that he could "rise up" in the Russian-owned Gazprom oil company, if he worked for them.

Trump aide Carter Page, who has collaborated with Russians, has helped define Trump's energy policy.

Politico reports that prior to that time, Page "spent five years in the Navy and served as a Marine intelligence officer in the western Sahara. During his Navy days, he spent lavishly and drove a black Mercedes... What appeared to [later] recommend him to Trump was his boundless enthusiasm for Putin and his corresponding loathing of Obama and [Hillary] Clinton. Page's view of the world was not unlike the Kremlin's. Boiled down: the United States' attempts to spread democracy had brought chaos and disaster."

The *Washington Post* says that Page "has praised Putin as a better leader than former President Barack Obama."

The Special Prosecutor concluded that the Russians' interest in Mr. Page was because he was close to Donald Trump. But in 2017 President Trump claimed Page was a "low-level" advisor "who nobody even knew," despite the fact that he himself had specifically mentioned Carter Page by name in a March 21, 2016 interview with the *Washington Post* editorial board and cited him as one of his top advisers.

Carter Page had far more positive views of Putin's regime than most Americans, and he was opposed to U.S. sanctions against Russia. He sent at least one email applauding the Trump campaign's success at getting a plank removed from the Republican party platform that even GOP Senators had been demanding, which called for "providing lethal defensive weapons" to Ukraine to combat Russian incursions.

Although the Special Prosecutor did not find that Page "coordinated with the Russian government in its efforts to interfere in the 2016 Presidential election," the Mueller report says investigators were "unable to obtain additional evidence or testimony about who Page may have met or communicated with in Moscow; thus, Page's activities in Russia—as described in his emails with the campaign—were not fully explained." So that does not mean exoneration per se; it could mean there just wasn't enough evidence to charge him with a crime...yet.

The second man who was tapped to advise Donald Trump on energy policy, George Papadopoulos, was suggested to the campaign by an Iowa radio talk show host, Sam Clovis. The Mueller report reveals that "Clovis performed a Google search on Papadopoulos, learned that he had worked at the Hudson Institute, and believed that he had credibility on energy issues." In one of their first conversations, Clovis told Papadopoulos that Russia policy would be an important focus for the campaign, the Mueller investigation alleges.

Ominously for Donald Trump, Mr. Papadopoulos apparently revealed over drinks with a top Australian diplomat to the United Kingdom, that he had been told that Russia was in possession of emails relating to Hillary Clinton.

The Mueller report states that "Papadopoulos had suggested to a representative of that foreign government that the Trump campaign had received indications from the Russian government that it could assist the campaign through the anonymous release of information damaging to Democratic Presidential candidate Hillary Clinton. That information prompted the FBI on July 31, 2016, to open an investigation into whether individuals associated with the Trump campaign were coordinating with the Russian government in its interference activities."

Unlike Carter Page, Mr. Papadopoulos pleaded guilty to lying to the FBI and served a brief term in jail.

The question still lingers: why would any candidate for President of the United States surround himself with so many people who had ties to Russia and authoritarian Vladimir Putin, and rely on them for policy advice? The Mueller report claims there are at least 127 known points of contact between people associated with the Trump

Best buds? This Photoshopped image of Putin and Trump is unfortunately closer to truth than parody.

campaign and Russian government-linked individuals or entities from November of 2015 all the way up to January of 2017.

Since he's become President, Donald Trump has had secret meetings with Vladimir Putin at which he has insisted no other U.S. officials be present—just a translator, sworn to secrecy. What's up with *that*? This is totally unprecedented and has been widely criticized even by members of Trump's own national security team. And everyone has probably seen the President of the United States standing on stage next to Mr. Putin and saying he believes Russia's denials that it interfered in our 2016 election, despite the unanimous view by U.S. intelligence agencies that Russia *did*.

Donald Trump's financial history with Russian oligarchs and at least $3 billion in loans goes back several decades, when his multiple, massive business failures and bankruptcies (over $1.17 billion in just one decade) convinced U.S. banks not to loan Trump any more money. So he sought funding from foreign sources.

The Russian government wants to prevent independent oil and gas exploration in order to solidify dependence by various countries on imported Russian gas. NATO's Secretary General Anders Rasmussen confirms it. Oil and gas represent about 50 percent of Russia's exports. A *Washington Examiner* story claimed that some prominent

American environmental groups were knowingly or unknowingly receiving money from Russia to oppose fracking.*

That seems plausible, considering the fact that with the U.S. far less reliant on foreign sources of oil, our country has actually been able to become the third largest *exporter* of oil—behind Russia and Saudi Arabia. So we compete on the world market with Russia. Therefore, Putin has an interest in supporting groups which oppose fracking, which is the newer method of extracting oil from shale in our country.

That does not mean, however, that green groups which oppose harmful practices such as fracking (which produces dirtier oil, by the way, than conventional oil wells), don't sincerely want to prevent environmental harm. It might surprise these groups—and their supporters—to know Putin is on their side, but for entirely different reasons.

Russia wants to subvert anything that constitutes energy efficiency so it's likely that country has attempted to influence the Trump administration to curtail support for clean energy alternatives like solar and wind.

Like on so many other issues, Vladimir Putin appears to be a consigliere for Donald Trump's environmental policies.

At first glance it may seem to be a contradiction that on the one hand Russia wants to discourage energy independence on the part of various nations around the world, and yet it is supporting Donald Trump who wants to bolster the coal industry.

However, coal is not a commodity that Russia exports to the United States anyway. But since cheaper natural gas and renewable energy sources decreased the demand for coal as a fuel for electricity generation, we are able to export it to dozens of countries around the world. The largest markets for us are India (12.8 percent of all exports), Netherlands (12 percent), Japan (8.5 percent), South Korea (8.5 percent), and Brazil (6.7 percent).

Russia is a major producer of coal and last year its export of coal worldwide hit a record high. Russian producers continue to seek domination on the European market and have been putting in a lot of effort to grab more market shares on the Asian markets such as South Korea and Taiwan. *So Putin would rather see America burn its supplies of*

---

*Fracking is the process of injecting liquid at high pressure into subterranean rocks, boreholes, etc. so as to force open existing fissures and extract oil or gas. In addition to air and water pollution, fracking also increases the potential for oil spills, which can harm the soil and surrounding vegetation. Fracking may also cause earthquakes due to the high pressure used to extract oil and gas from rock and the storage of excess wastewater on site.

coal domestically than pose competition to Russia by selling it on the world market.

Is it a coincidence Donald Trump is promoting domestic consumption of coal? That's exactly what Putin wants.

It's also revealing that Trump is in favor of resuming an industry the government itself says is ripping off taxpayers. The U.S. public is effectively subsidizing coal companies—to the tune of $28.9 billion over the past 30 years.

In another effort to prop up the coal industry, the Department of Energy has decided to force electric utilities to buy power from coal-fired power plants. In this case, the rationale they've manufactured centers around a bogus argument about national security and grid reliability.

Former California Republican Gov. Arnold Schwarzenegger made a video uploaded to Facebook in which he references a popular movie role he played to mock what he describes as efforts to "save an industry that is poisoning the environment."

"So President Trump, I know you really want to be an action hero, right?" Schwarzenegger says, while looking at a Trump bobble-head doll. "So take it from the Terminator, you're only supposed to go back in time to protect future generations. But your administration attempts to go back in time to rescue the coal industry, which is actually a threat to future generations."

He then compares Trump's attempt to "rescue the coal industry" to rescuing other relics from America's past.

"It is foolish to bring back laughable, outdated technology to suit your political agenda," Schwarzenegger says. "I mean, what are you going to bring back next? Floppy disks? Fax machines? Beanie Babies? Beepers? Or Blockbuster? Think about it. What if you tried to save Blockbuster?"

Herman Schwartz of *The Nation* magazine writes that "In its zeal to enhance the profits of fossil-fuel, chemical, and other industrial polluters, this administration is exposing many millions of American children to dangerous and often life-threatening poisons and contaminants.

"Lead, asbestos, poisonous insecticides, fossil-fuel emissions, and many other toxic pollutants contaminate our air, water, food, and homes. Children are much more vulnerable to these toxic substances than adults. Their central-nervous, immune, and other systems are still undeveloped, and exposure to toxic substances can cause irreversible damage; a child's lungs are particularly sensitive. Even a fetus is at risk if a woman is exposed to toxins during pregnancy.

"Our protection against these threats depends on the Environmental Protection Agency's enforcement of the Clean Water and Clean Air Acts, and other anti-pollution laws. For all their foot-dragging and other faults, prior Democratic and Republican administrations did enforce these laws, significantly reducing some of the worst air and water hazards.

"No more. Trump's entire administration is packed with sworn enemies of the laws they are supposed to administer, but nowhere more so than at the EPA."

Although as a candidate, Donald Trump promised to "drain the swamp" and limit the influence of lobbyists over the federal government, his administration has done the opposite. In his first two years, the Trump administration hired at least 164 former lobbyists in a wide variety of influential positions, according to an analysis by the nonprofit Center for Responsive Politics.

At the EPA, nearly half of the political appointees hired by the Trump administration have strong ties to industries regulated by the agency, according to research by the Associated Press. About a third of these EPA appointees—including the current administrator, Andrew Wheeler—formerly worked as registered lobbyists or lawyers for fossil fuel companies, chemical manufacturers, or other corporate clients.

As Attorney General of Oklahoma, Trump's first EPA Administrator, Scott Pruitt, challenged EPA regulations in court 14 times. Pruitt hired former Oklahoma banker Albert Kelly to head the Superfund program, which is responsible for cleaning up the nation's most

Cost of new EPA Coal Rules: Up to 1,400 more deaths a year — *The New York Times*

contaminated land. Kelly completely lacked any experience with environmental issues, and had just received a lifetime ban from working in banking, his career until then, due to "unfitness to serve."

The Nation reports that "At the EPA for nearly 17 months, Pruitt turned the agency over to fossil-fuel and other industry lobbyists, both within the agency and outside. Pruitt's nonstop scandals became so embarrassing that he was forced to resign, leaving Deputy Administrator Andrew Wheeler as acting administrator."

For nearly two decades, "Wheeler was a chief counsel for Oklahoma Senator James Inhofe, the Senate's most vehement climate-change denier. Between 2009 and 2017 Wheeler was a prominent lobbyist who represented many clients with business before the federal government, including the Domestic Fuels Solution Group, chemical manufacturer Celanese Corporation, and biodiesel producer Darling Ingredients; his main client was coal baron Robert E. Murray."

Criminal enforcement by the EPA is at a 30-year low under Trump. Neither Trump nor the department heads he has appointed believe carbon dioxide is a primary contributor to global warming.

To head the Department of Energy, Trump picked former Texas Governor and failed Presidential candidate Rick Perry—who had previously advocated abolishing the agency.

Ryan Zinke was initially appointed to head the Department of the Interior. Zinke has a history of opposing environmental groups, maintaining a lifetime score of four percent with the League of Conservation Voters. Several environmental groups spoke out against Trump's choice of Zinke. "Zinke embodies the worst kind of magical thinking in Congress: that government welfare handouts can save dying coal companies and crumbling oil and gas giants," Diana Best of Greenpeace said. Ultimately, Zinke got bogged down in ethics investigations that troubled even the corrupt Trump administration, and was forced to leave.

In January 2019, Zinke was replaced with David Bernhardt, who as Zinke's previous assistant helped the department embark on a program of deregulation and substantially increased fossil fuel sales on public land. In March 2019, *Politico* reported that heads of the oil industry lobbyist group Independent Petroleum Association of America (IPAA) boasted about their ties to Bernhardt. After an uproar from Democrats over his conflicts of interest and a prior career as an oil and gas lobbyist—Bernhardt has faced a wave of scandals that has prompted the Interior Department's watchdog to launch an investigation of Bernhardt over ethics concerns, along with a half-dozen other

Los Angeles smog before air quality standards were enacted.

DOI senior officials.

To head the Council on Environmental Quality, Trump chose Kathleen Hartnett White, who called renewable energy "unreliable and parasitic" and suggested that climate regulation is "a conspiracy pushed by communists." Her nomination was withdrawn in February 2018 as she did not garner enough support in the Senate, even among Republicans who normally rubberstamp the President's nominees.

Donald Trump announced what he called The America First Energy Plan but it does not mention renewable energy and instead reflects the President's focus on fossil fuels.

A *New York Times* analysis, based on research from Harvard Law School, Columbia Law School and other sources, counts 78 environmental rules already nullified or on the way out under Mr. Trump.

The harmful impact of this cannot be overstated.

Trump has taken the anti-regulation fervor of past GOP administrations and far exceeded anything Reagan or the Bushes attempted to do. He would turn back the clock to when Los Angeles smog was so bad that on some days, the air was so polluted that parents kept their kids out of school, athletes trained indoors, citrus growers and sugar-beet producers watched in dismay as their crops withered, the elderly and young crowded into doctors' offices and hospital Ers with throbbing heads and shortness of breath.

In 1943, the first big smog scare sent residents running from what they assumed was a Japanese gas attack. The city's once clear coastal air had become a tear-inducing haze, and no one knew what was causing it.

Diesel buses arrived on the scene in the 1940s but were not understood to be a major source of air pollution that they were. It's no coincidence that the presence of air pollution became a problem in the 1940s, when the number of cars in L.A. had doubled from one to two million.

In the 1950s, automobile exhaust became a prime suspect. Even before they were crisscrossed by clogged highways, the air in Southern California's inland valleys was subject to pollution. Haze, possibly caused by the use of smudge pots to prevent frost on the Pomona Valley's citrus trees, would partially obscure snow-capped Mount Baldy.

Urban air pollution is often seen as an unfortunate but inevitable byproduct of industrialization.

In 1948, a Caltech biochemist named Arie Haagen-Smit made the connection with car exhaust. Even after his discovery, Haagen-Smit had to fight the oil industry-backed researchers who attempted to disprove his ideas.

Congress passed the Clean Air Act in 1963. A little over a decade later, national laws requiring catalytic converters when new automobiles were introduced. The new laws helped roll back the L.A. haze, but the legislation came too late for the millions of people who had grown up under oppressive smog. By 1987, an estimated 27 percent of Angelenos were living with "severely damaged" lungs.

Today, ozone levels in L.A. are 40 percent of what they were in 1970—and that's with double the cars.

Though smog is still a problem in the California metropolis, KCET television reports that "air pollution rarely cripples the city in present times as it did in the mid-20th century. Severe smog has largely abated...thanks to citizen activism, scientific advances, and landmark environmental legislation that allowed the EPA to regulate air pollutants. That leaves many Southern Californians with only hazy memories of severe smog."

The New York Times reports, "All told, the Trump administration's environmental rollbacks could lead to at least 80,000 extra deaths per decade and cause respiratory problems for more than one million people, according to a separate analysis conducted by researchers from Harvard. That number, however, is likely to be 'a major underestimate of the global public health impact,' said Francesca Dominici, a professor of biostatistics at the Harvard School of Public Health."

How Trump is rolling back environmental protections

Here is a list *The Times* compiled of the multitude of ways the Trump administration is methodically trying to take actions that will relieve major polluters from public accountability and regulation that has had bipartisan support for decades:

1. Canceled a requirement for oil and gas companies to report methane emissions.

2. Revised and partially repealed an Obama-era rule limiting methane emissions on public lands, including intentional venting and flaring from drilling operations.

3. Loosened a Clinton-era rule designed to limit toxic emissions from major industrial polluters.

4. Stopped enforcing a 2015 rule that prohibited the use of hydrofluorocarbons, powerful greenhouse gases, in air-conditioners and refrigerators.

5. Repealed a requirement that state and regional authorities track tailpipe emissions from vehicles traveling on federal highways.

6. Amended rules that govern how refineries monitor pollution in surrounding communities.

7. Directed agencies to stop using an Obama-era calculation of the "social cost of carbon" that rulemakers used to estimate the long-term economic benefits of reducing carbon dioxide emissions.

8. Withdrew guidance that federal agencies include greenhouse gas emissions in environmental reviews. (But several district courts have ruled that emissions must be included in such reviews.)

9. Reverted to a weaker 2009 pollution permitting program for new power plants and expansions.

10. Proposed weakening Obama-era fuel-economy standards for cars and light trucks. The proposal also challenges California's right to set its own more stringent standards, which can be followed by other states.

11. Announced intent to withdraw the United States from the Paris climate agreement (but the process of withdrawing cannot be completed until 2020).

12. Proposed repeal of the Clean Power Plan, which would have set strict limits on carbon emissions from coal- and gas-fired power plants. In August 2018, the EPA drafted a replacement plan, called the Affordable Clean Energy rule, that would let states set their own rules—a prescription for disaster in GOP-lead states.

13. Proposed eliminating Obama-era restrictions that in effect required newly built coal power plants to capture carbon dioxide emissions.

14. Proposed a legal justification for weakening an Obama-era rule that limited mercury emissions from coal power plants.

15. Proposed revisions to standards for carbon dioxide emissions from new, modified and reconstructed power plants.

16. Began review of emissions rules for power plant start-ups, shutdowns and malfunctions. In September 2018, EPA officials said they were considering repealing the rule.

17. Proposed relaxing Obama-era requirements that companies monitor and repair methane leaks at oil and gas facilities.

18. Proposed changing rules aimed at cutting methane emissions from landfills.

19. Announced rewrite of an Obama-era rule meant to reduce air pollution in National Parks and wilderness areas.

20. Weakened oversight of some state plans for reducing air pollution in national parks. (In Texas, the EPA rejected an Obama-era plan that would have required the installation of equipment at some coal-burning power plants to reduce sulfur dioxide emissions.)

21. Proposed repealing leak-repair, maintenance and reporting requirements for large refrigeration and air-conditioning systems containing hydrofluorocarbons.

22. Lifted a freeze on new coal leases on public lands.

23. Repealed an Obama-era rule governing royalties for oil, gas and coal leases on federal lands, which replaced a 1980s rule that critics said allowed companies to underpay the federal government.

24. Made significant cuts to the borders of two national monuments in Utah and recommended border and resource management changes to several more.

25. Revoked an Obama-era executive order designed to preserve ocean, coastal and Great Lakes waters in favor of a policy focused on energy production and economic growth.

26. Rescinded water pollution regulations for fracking on federal and Indian lands.

27. Scrapped a proposed rule that required mines to prove they could pay to clean up future pollution.

28. Withdrew a requirement that Gulf oil rig owners prove they could cover the costs of removing rigs once they have stopped producing.

29. Approved construction of the Dakota Access pipeline less

than a mile from the Standing Rock Sioux reservation. Under the Obama administration, the Army Corps of Engineers had said it would explore alternative routes.

30. Changed how the Federal Energy Regulatory Commission considers the indirect effects of greenhouse gas emissions in environmental reviews of pipelines.

31. Permitted the use of seismic air guns for gas and oil exploration in the Atlantic Ocean. The practice, which can kill marine life and disrupt fisheries over very broad areas of many miles, was blocked under the Obama administration.

32. Proposed opening most of America's coastal waters to offshore oil and gas drilling.

33. Expedited an environmental review process to clear the way for drilling in the Arctic National Wildlife Refuge.

34. Ordered review of regulations on oil and gas drilling in national parks where mineral rights are privately owned.

35. Proposed changes to regulations for oil well control and blowout prevention systems implemented after the 2010 Deepwater Horizon explosion and oil spill.

36. Recommended shrinking three marine protected areas, or opening them to commercial fishing.

37. Proposed revisions to regulations on offshore oil and gas exploration by floating vessels in the Arctic that were developed after a 2013 accident.

38. Revoked Obama-era flood standards for federal infrastructure projects, like roads and bridges. The standards required the government to account for sea-level rise and other climate change effects.

39. Relaxed the environmental review process for federal infrastructure projects.

40. Revoked a directive for federal agencies to minimize impacts on water, wildlife, land and other natural resources when approving development projects.

41. Revoked a 2016 order promoting "climate resilience" in the northern Bering Sea region in Alaska.

42. Revoked an Obama-era order that had set a goal of cutting the federal government's greenhouse gas emissions by 40 percent over ten years.

43. Reversed an update to the Bureau of Land Management's public land use planning process.

44. Withdrew an Obama-era order to consider climate change in managing natural resources in National Parks.

45. Restricted most Interior Department environmental studies to one year in length and a maximum of 150 pages, citing the need to reduce paperwork.

46. Withdrew a number of Obama-era Interior Department climate change and conservation policies that the agency said could "burden the development or utilization of domestically produced energy resources."

47. Eliminated the use of an Obama-era planning system designed to minimize harm from oil and gas activity on sensitive landscapes, such as national parks.

48. Eased the environmental review processes for small wireless infrastructure projects with the goal of expanding 5G wireless networks.

49. Announced plans to speed up and streamline the environmental review process for forest restoration projects.

50. Overturned a ban on the use of lead ammunition and fishing tackle on federal lands. This is killing birds like bald eagles.

51. Overturned a ban on the hunting of predators in Alaskan wildlife refuges.

52. Ended an Obama-era rule barring hunters on some Alaska public lands from using bait to lure and kill grizzly bears.

53. Withdrew proposed limits on endangered marine mammals and sea turtles unintentionally caught by fishing nets on the West Coast.

54. Amended fishing regulations for a number of species to allow for longer seasons and higher catch rates.

55. Rolled back an Obama-era policy aimed at protecting migratory birds.

56. Overturned a ban on using parts of migratory birds in handicrafts made by Alaskan Natives.

57. Proposed stripping the Endangered Species Act of key provisions.

58. Released a plan to open nine million acres of Western land to oil and gas drilling by weakening habitat protections for the sage grouse, an imperiled bird with an elaborate mating dance.

59. Narrowed the scope of a 2016 law mandating safety assessments for potentially toxic chemicals, like dry-cleaning solvents and paint strippers. The EPA will focus on direct exposure and exclude air, water and ground contamination.

60. Reversed an Obama-era rule that required braking system upgrades for "high hazard" trains hauling flammable liquids, like oil

and ethanol.

61. Removed copper filter cake, an electronics manufacturing byproduct comprised of heavy metals, from the "hazardous waste" list.

62. Rejected a proposed ban on chlorpyrifos, a potentially neurotoxic pesticide. In August 2018, a federal court ordered the E.P.A. to ban the pesticide, but the agency appealed the ruling.

63. Proposed eliminating a program designed to limit exposure to lead, which is known to damage brain and nervous system development.

64. Announced a review of an Obama-era rule lowering coal dust limits in mines. The head of the Mine Safety and Health Administration said there were no immediate plans to change the dust limit, but the review is continuing.

65. Revoked a rule that prevented coal companies from dumping mining debris into local streams.

66. Withdrew a proposed rule reducing pollutants, including air pollution, at sewage treatment plants.

67. Revoked federal rules regulating coal ash waste from power plants and granted oversight to the states.

68. Withdrew a proposed rule requiring groundwater protections for certain uranium mines.

69. Proposed rolling back protections for certain tributaries and wetlands that the Obama administration wanted covered by the Clean Water Act.

70. Delayed by two years an EPA rule regulating limits on toxic discharge, which can include mercury, from power plants into public waterways.

71. Prohibited funding environmental and community development projects through corporate settlements of federal lawsuits.

72. Announced intent to stop payments to the Green Climate Fund, a United Nations program to help poorer countries reduce carbon emissions.

73. Reversed restrictions on the sale of plastic water bottles in national parks designed to cut down on litter, despite a Park Service report that the effort worked.

74. Proposed limiting the studies used by the EPA for rulemaking to only those that make data publicly available. (The move was widely criticized by scientists, who said it would effectively block the agency from considering landmark research that relies on confidential health data.)

75. Proposed changes to the way cost-benefit analyses are con-

ducted under the Clean Air Act, Clean Water Act and other environmental statutes.

76. Delayed compliance dates for federal building efficiency standards until Sept. 30, 2017. No updates have been published, and the status of the rule remains unclear.

77. Proposed withdrawing efficiency standards for residential furnaces and commercial water heaters designed to reduce energy use.

78. Withdrew a proposed rule that would inform car owners about fuel-efficient replacement tires. The Transportation Department is scheduled to republish a proposal in June 2019.

As *The New York Times* notes, "The process of rolling back regulations has not always been smooth. In some cases, the administration has failed to provide a strong legal argument in favor of proposed changes or agencies have skipped key steps in the rulemaking process, like notifying the public and asking for comment. In several cases, courts have ordered agencies to enforce their own rules."

A case in point was in early April 2019 when a federal judge struck down one of President Donald Trump's orders, and restored an Obama-era ban on offshore drilling in certain parts of the Arctic Ocean and dozens of canyons in the Atlantic Ocean. The bans were a key part of Obama's environmental legacy.

Presidents have the power under a federal law to remove certain lands from development but cannot revoke those removals, according to U.S. District Court Judge Sharon Gleason.

She ruled that Trump exceeded his authority, and threw out his executive orders he issued in this regard.

"The wording of President Obama's 2015 and 2016 withdrawals indicates that he intended them to extend indefinitely, and therefore be revocable only by an act of Congress," said Judge Gleason.

Erik Grafe, an attorney with Earthjustice, welcomed the ruling, saying it "shows that the President cannot just trample on the Constitution to do the bidding of his cronies in the fossil fuel industry at the expense of our oceans, wildlife and climate."

Earthjustice represented numerous environmental groups that sued the Trump administration over the April 2017 executive order reversing the drilling bans. At issue in the case was the Outer Continental Shelf Lands Act.

In the Atlantic, Obama banned exploration in 5,937 square miles of underwater canyon complexes, citing their importance for marine mammals, deep-water corals, valuable fish populations and migratory whales.

In other defeats for the Trump administration, the courts performed their crucial role as part of the checks-and-balances system under our Constitution, even as Congress is derelict in this regard.

So Trump was forced by lawsuits to reinstate a rule aimed at improving safety at facilities that use hazardous chemicals following a federal court order.

He approved the Keystone XL pipeline rejected by President Barack Obama, but a federal judge blocked the project from going forward, saying the Trump administration did not present a "reasoned explanation" for the approval.

Trump delayed a compliance deadline for new national ozone pollution standards by one year, but later reversed course.

He was forced to suspend an effort to lift restrictions on mining in Bristol Bay, Alaska.

Trump announced his intention to regulate paint removers containing methylene chloride after pressure from families who had lost relatives to poisoning. The EPA had indicated it would not finalize regulatory action on the substance. (But so far it has not moved to ban its commercial use.)

The administration delayed implementation of a rule regulating the certification and training of pesticide applicators, but a judge ruled that the EPA had done so illegally and declared the rule in effect.

Trump initially delayed publishing efficiency standards for household appliances, but later published them after multiple states and environmental groups sued.

He reissued a rule limiting the discharge of mercury by dental offices into municipal sewers after a lawsuit by the Natural Resources Defense Council, an advocacy group.

The White House re-posted a proposed rule limiting greenhouse gas emissions from aircraft, after initially changing its status to "inactive" on the EPA website.

Trump removed the Yellowstone grizzly bear from the Endangered Species List, but the protections were later reinstated by a federal judge.

Thank God for the courts! But this list does not include new rules proposed by the Trump administration that do not roll back previous policies, nor does it include court actions that have affected environmental policies independent of executive or legislative action.

And the Administration's flagrant disregard of court rulings in clear violation of the U.S. Constitution's separation of powers connotes an unprecedented lawlessness that alone are grounds for impeachment of President Donald Trump and removal from office.

Another ecological disaster compounded by Trump neglect

Besides the debacle at our southern border in which immigrants are being badly abused by the Trump administration, there has been a second humanitarian crisis so far that reflects the President's vileness, lack of compassion and his incompetence.

Hurricane Irma struck the Carribbean in August 2017. It was an extremely powerful and catastrophic Cape Verde hurricane, the strongest observed in the Atlantic in terms of maximum sustained winds since Wilma in 2005, and the strongest storm on record to exist in the open Atlantic region.

Once again, scientists linked the intensity and rapid development of this storm to global warming.

Irma was followed two weeks later by Hurricane Maria—a Category 5 storm—that devastated Dominica, the U.S. Virgin Islands, and Puerto Rico—the latter home to 3.4 million people. It is regarded as the worst natural disaster on record to affect those islands and is also

A man who has lived in San Isidro, outside the Puerto Rico capital since he was nine years old, stands amidst the rubble in his neighborhood, torn apart by Hurricane Maria, on Sept. 28, 2017.

one of the deadliest Atlantic hurricanes ever.

As *Time* magazine chronicled, "The fifth strongest storm ever to strike the U.S. hit Puerto Rico on Sept. 20 with stronger winds than Irma brought to Florida and the kind of rain that Hurricane Harvey dumped on Houston. It made landfall on a Wednesday, and in the digital age, its effects were well documented by Friday: parts of San Juan, the capital of this U.S. territory, were underwater. The verdant island was stripped of its foliage. U.S. citizens lapped water from natural springs. But on the mainland, the topics of the day were a special election in Alabama, the latest GOP stab at repealing Obamacare and a fight President Donald Trump had picked with the NFL."

When Maria struck, approximately 80,000 Pureto Ricans remained without power due to Irma. Much of the housing and infrastructure were left beyond repair, while the island's lush vegetation was practically eradicated. The islands of Guadeloupe and Martinique endured widespread flooding, damaged roofs, and uprooted trees. Puerto Rico also suffered a major humanitarian crisis that was compounded by a lack of resources and the slow relief process.

Donald Trump seemed not to know Puerto Rico was a U.S. Territory and as such, his responsibility to protect.

The *National Geographic* described the situation in Puerto Rico thusly: "With no power, running water was cut off for much of the population. Communications to and from Puerto Rico were nearly impossible for days. Airports were shut down, delaying recovery efforts, since supplies had to be airlifted or shipped in. And the Federal Emergency Management Agency, charged with disaster relief, was already stretched thin after historic storms earlier [that] summer in Texas and Florida.

"The result was the longest major power outage in U.S. history, and many communities on the island were left without running water for months. Toilets couldn't flush; there was no water for showers, baths, or washing clothes. People had to rely on bottled water, but supplies were limited. Useless electric stoves had to be replaced with propane ones. Without refrigeration, food rotted and vital medicines spoiled. Only those with gas-powered generators could ward off darkness after dusk—for a few brief hours. Forget about air conditioners to relieve the sweltering heat. All the modern conveniences we take for granted were left behind."

Puerto Rico and surrounding islands practically reverted to the Stone Age. Meanwhile, Donald Trump turned a blind eye toward the suffering or—when he spoke about it—villified the people there.

It was an insulting and meaningless gesture: President Trump tossing paper towels at the victims of hurricane Maria in Puerto Rico who were ravaged by the storm.

When he visited Puerto Rico, he was photographed ludicrously throwing rolls of paper towels to storm victims, in a jovial manner, as if their plight was all a big joke to him.

A year later, thousands of Puerto Ricans were still without power—an unimaginable condition and totally inexcusable.

The President is a lot less concerned about its fate. *National Geographic* magazine says that "Even after power and water are restored across the the island, people will still be dealing with the storm's effects. 'The storm takes away the foundations of society. Everything you thought gave you certainty is gone,' says psychologist Domingo Marqués, 39, an associate professor at Albizu University in San Juan. 'You see people anxious, depressed, scared.' Marqués estimates that 30 to 50 percent of the population is experiencing post-traumatic stress disorder, depression, or anxiety."

Donald Trump insists his administration did a "fantastic job" in responding to the crisis and that he is "the best thing that ever happened to Puerto Rico." He has tweeted and said at a May 2019 rally in Florida that Puerto Rico got a record $91 billion in recovery aid (while Floridians wait for hurricane help). The truth is, only $11.2 billion has been spent—an amount that is grossly inadequate.

PolitiFact asserts that "His comments about hurricane aid follow other misleading statements he made related to the 2017 hurricanes

Irma and Maria in Puerto Rico. Trump mischaracterized the death toll (thought to be around 6,000 people), saying the estimate was 'done by Democrats,' and he exaggerated the poor condition of Puerto Rico's electrical grid before the hurricanes."

In January 2019, the Trump White House opposed supplemental funds, even though food aid for the island's poor was about to run out. *Politico* reported that "At the same time, billions of dollars in community development appropriations have yet to leave Washington —a year after being approved by Congress to assist in the recoveries from hurricanes Maria and Irma."

Once again—just like he has done when demonizing refugees at our southern border—Trump has shown his hatred of low-income, often Latino families, even though Puerto Ricans are U.S. citizens.

A vindictive and petty man, according to *Politico*, "Donald Trump the President remains furious with Puerto Rico Gov. Ricardo Rossello for backing Sen. Bill Nelson (D-Florida) who only narrowly lost to Trump's choice, former Republican Gov. Rick Scott, in the 2018 midterms. "The driver of the new heightened toxicity is the election, in Rossello endorsing Bill Nelson," said one close observer. "That was it. You have a President basically going at it like we were in a Third World country."

Can you imagine any prior President of the United States intentionally harming thousands of victims of a natural disaster that occurred on U.S. soil, simply as payback for what he perceives as political disloyalty on the part of one office holder whom he hates?

Worsening Puerto Rico's problems is that the island's poor are already disadvantaged because they don't receive the same food stamp benefits as other U.S. citizens, including residents of the Virgin Islands, only about 50 miles east in the Caribbean. That's because during budget cuts made by President Ronald Reagan in the 1980s, Puerto Rico was made a sacrificial lamb. Beginning in 1982, the island was cut out of the food stamp entitlement, now called the Supplemental Nutrition Assistance Program (SNAP).

Though Congress eventually provided a block grant, it is far lower than needed. *Politico* notes that "Thousands of families below the poverty line are left uncovered, all at a time while the Agriculture Department's own reports show rising food costs in Puerto Rico. Even before Maria, for example, it was estimated that the average household on Puerto Rico was spending nine percent more on food in 2013 than the average American household."

One wonders: if the President's grandiose estate, the 128-room,

62,500-square-foot Mar-a-Lago mansion in South Florida were struck by a similar storm, how long do you suppose it would go without electricity or running water?

In fact, Trump received a $17 million insurance payment for hurricane damage to Mar-a-Lago after the 2005 Atlantic hurricane season, for damage to the "landscaping, roofing, walls, painting, leaks, artwork in the—you know, the great tapestries, tiles, Spanish tiles, the beach, the erosion," as he described it. However, Trump's former butler at the resort and later its "in house historian," disputed Trump's assertions (which seems to corroborate his former attorney, Michael Cohen's, testimony that Trump has often committed insurance and bank fraud). Some trees behind the resort had been flattened and some roof tiles were lost, but, said his former butler, "That house has never been seriously damaged. I was there for all of [the hurricanes]."

If there is any silver lining to the storm clouds Donald Trump has brought to our nation, it may be that membership of green groups has surged since his election. Attorneys general from 15 states have promised to defend the clean power plan. More than 600 multinationals from Starbucks to DuPont called on the incoming President to uphold the Paris Agreement. It would appear that across political lines, a majority of U.S. citizens look with disdain on Trump's behavior and policies, and that does not bode well for his re-election in 2020.

Can anyone imagine that if President Trump's $100 million, sprawling seaside estate, Mar-a-Lago in Palm Beach, Florida was struck by a hurricane like Puerto Rico, that it would be left without electricity or running water for more than a year?

6.
Our Vanishing Wilderness

WHERE DO YOU GO WHEN YOU WANT TO GET AWAY? Over 331 million Americans retreat to one of our National Parks, to explore some of the last vestiges of the natural environment left amidst the urban sprawl. But decades of neglect and underfunding have left our parks in desperate straits. Infrastructure is crumbling, budget cutbacks have reduced staff, and privatization has been a disaster. Drug smugglers, poachers, criminal thugs and illegal immigrants have invaded our formerly pristine wilderness, threatening visitors and rangers and leaving behind debris and human waste.

Yet Americans value their wild heritage as much as ever. Polling has shown that about 90 percent of voters nationwide support permanent public land protection (while 69 percent oppose measures to prevent it). Even in what seems to be an unusually political and polarized age, the value of our wilderness areas is all but universal. In a survey conducted after the 2016 election, most Trump voters said they oppose efforts to privatize or sell off public lands. Millions of Americans submitted comments to the Trump administration opposing its punitive review of national monument lands.

Regardless of public opinion on these matters, Donald Trump has aggressively pushed for more fossil fuel development across the United States by opening up new lands to drilling and mining, attempting to bail out money-losing coal power plants, and dialing back environmental regulations that stand in the way.

As previously mentioned, the Department of the Interior—in charge of mineral rights on federal lands, environmental stewardship, promoting outdoor recreation, and upholding treaty obligations with Native Americans—has been headed by a succession of disastrous political hacks and fossil fuel industry insiders nominated by Republican

Presidents since Ronald Reagan.

When Donald Trump took office, Ryan Zinke as Interior Secretary, presided over the largest rollback of federal land protections in U.S. history and opened up nearly all our coastal waters to oil drilling.

His successor, David Bernhardt, formerly worked at the Interior Dept. during the disastrous George W. Bush administration. Then he became an attorney representing oil companies. *Vox* reports that while working as Zinke's underling, "Bernhardt led reviews of key environmental regulations like the Endangered Species Act to make energy development easier on federal land. He also signed a secretarial order that undid agency rules for reducing the environmental impacts of activities on federal lands. This included best practices for mitigating climate change. Bernhardt has also made it harder to incorporate the latest science to guide Interior's policies."

"His years of lobbying on behalf of clients who stand to profit from Interior policy decisions are cause for serious concern," said Rep. Raúl Grijalva (D-AZ), the new chair of the House Natural Resources Committee, which has oversight duties for the Interior Department.

Besides the administrative assaults on our wilderness areas and the nation's other environmental assets, there is a legislative attack as well.

In a clever ruse engineered by Republicans and even supported by some Democrats, legislation has been drafted that preserves federal ownership of certain public lands, but offers for the first time leasing oil and gas drilling rights on them to energy companies who bid for the rights at bargain prices. Wilderness areas and our National Parks are also being opened up to trophy hunters who are killing off bears, elk, wolves, bison and other animals.

All of this is occurring in the context of our wilderness areas and National parks urgently needing an infusion of cash to make long-overdue repairs and maintenance to preserve them.

From Yellowstone to the Smoky Mountains, our parks are suffering from increased pollution and destructive uses, and the promotion of private interests at the expense of our treasured public heritage.

As of 2019, there is a staggering $11.6 billion maintenance backlog in National Park Service locales around the country. Yellowstone National Park needs over $800 million to address its long-neglected repair issues. Not to mention that thermal basins change constantly and boardwalks need to be moved.

The maintenance backlog for Olympic National Park and Mount Rainier is estimated to cost $121 million, putting it in the upper

Americans want to be able to access the scenic wonders of our National Parks.

range of parks with repair and rehabilitation projects that have been delayed. Gateway Recreation Area in New York needs $677 million for repairs.

Dana Soehn, spokesperson for the Great Smoky Mountains National Park, says park visitation has increased by 25 percent over the past decade, while staffing has decreased by 23 percent.

Grand Canyon National Park, which saw more than 6.2 million visitors in 2017, has over $329 million in deferred maintenance. That figure includes The Trans-Canyon Pipeline, which carries water to the South Rim and its visitor centers and hotels. The pipe's condition continues to deteriorate as it leaks and breaks.

As Rita Beamish of the *National Parks Traveler* points out, "The backlog is a result of budgets that haven't kept pace with the pressures of aging infrastructure and soaring park visitation, up from 256 million people across the parks in 1990, to 331 million last year." Congress after Congress has not been willing to spend the money we need to sustain world-class public lands.

So while Americans are flocking to our National Parks in record numbers, lawmakers have not adequately funded their preservation. And Donald Trump wants to strip additional money from the parks. (You gotta pay for that massive tax break for the super-rich and big corporations some way, ya know).

The Trump administration provided a very damaging budget blueprint for FY2018, asking Congress for draconian cuts to the Park Service. In a rare, bipartisan show of opposition, Congress refused the

Administration's proposed budget, which would have eliminated thousands of jobs at the agency, and instead provided it with *more* money than was originally proposed.

But Trump's 2019 budget blueprint called for a seven percent budget cut to the Park Service. This would result in a loss of nearly 2,000 ranger jobs. The proposal also includes specific cuts to cultural programs, land acquisition and the Centennial Challenge, a program that manages philanthropic donations, according to the National Parks Conservation Association.

"This budget is the Republican approach to governing in a nutshell: cut taxes for the super-rich and then, when it's time to fund national priorities, lecture us about tightening our belts," complained Rep. Raúl Grijalva. "If you think environment conservation is an unaffordable luxury, you'll love this plan. This isn't worth the paper it's printed on." Meanwhile, taxpayers have shelled out over $143 million for Trump to play golf at his various resorts.

At the National Trust for Historic Preservation, Thomas J. Cassidy called Trump's budget proposal devastating.

"Federal preservation programs and policies are essential to protecting places that tell our American story," Cassidy insisted. "They help ensure that historic communities remain engines of economic revitalization, and that families can learn and engage with our nation's

Red-orange-pink spires at Bryce Canyon National Park in southern Utah.

history in compelling ways. But these proposed funding cuts will impact every state and Congressional district in the country. Vital resources for protecting historic places—such as the Historic Preservation Fund and its support for state and tribal historic preservation offices, the Land and Water Conservation Fund, and National Heritage Areas—would by decimated by this proposal."

Our National Parks are supposed to receive special status

By law, our National Parks have been granted a status above the rest: America's most special places, where natural beauty and all its attendant pleasures—quiet waters, the scents of fir and balsam, the splendor of colorful wildflowers, the songs of myriad birds, and the dark of a night sky unsullied by city lights—are sacrosanct.

Historically, conservation has always been the top priority, trumping any suggested use that might degrade them. But the second Bush administration held a different view.

Ronald Reagan's antipathy toward nature was bad enough, and then George H.W. Bush was a disappointment. But it was his son, George Jr. who accelerated the Republican-led assault on the environment that has been embraced by Mr. Trump, despite an interlude when Barack Obama tried to help.

In 2000 alone, Bush and his Vice Presidential running mate, Dick Cheney, received more than $44 million in campaign contributions from such industries as mining, oil and gas, timber, coal, chemicals, and various other major sectors that harm the environment. According to a report published in *Newsweek* in 2000, the Bush campaign even promised contributors that their donations would receive tracking codes, so they could later be correctly identified and compensated through political decisions.

In 2000, candidate George W. Bush pledged to eliminate the maintenance backlog of our National Parks. But after he won re-election to a second term, he decided he only wanted the government to cover just 70 percent of the parks' payroll and utility costs in 2007. He planned $100 million in cuts for 2008. By that point, some parks were already reporting daily operating budget shortfalls in excess of 50 percent. That left them scrambling for charitable donations.

During Bush's eight years in office, the Bureau of Land Management, charged with the oversight of public land, was told that issuing new leases for oil and gas exploration was its highest—often its only—priority. Millions of acres—including some of the nation's most environmentally sensitive public lands—were opened up to logging, mining, and oil and gas drilling. A prime impetus for this change in policy was Vice President Dick Cheney, a Svengali character in American politics—and a villain to environmentalists for the harm Cheney helped inflict on our natural resources.

This despite the fact that Dick Cheney is an avid fisherman, and a hunter—though not a very good one, as he accidentally wounded a member of his hunting party by discharging his shotgun recklessly. The 78-year-old victim, Harry Whittington, suffered serious injuries including a collapsed lung, atrial fibrillation and a mild heart attack after he was sprayed with more than 200 birdshot pellets from Cheney's shotgun during a quail-hunting trip in 2006—and Cheney still had not apologized to the man, according to an article in *The Hill* in 2016. Mr. Whittington's voice was permanently affected by his injuries.

Cheney was the power behind the throne.

The exact circumstances surrounding the incident and the timetable remain a mystery as three different days have been stated for the mishap and there are conflicting accounts of what happened.

To fully understand the Bush assault on our National Parks, it's important to know something about his Vice President. Dick Cheney was originally chosen to chair the committee tasked with finding a running mate for Bush, but instead managed to get himself selected to be Bush's V.P. He never should have even been a heartbeat away from the Presidency since Cheney had a long history of cardiac problems, having his first of five heart attacks at age 37. He ultimately became the recipient—at taxpayer expense—of an artificial heart pump while he awaited a successful heart transplant, again at taxpayer expense.

Wil Hylton blames Cheney for the pre-emptive war in Iraq, writing that " despite overwhelming skepticism within the government of a link between Iraq and Al Qaeda—resulting in the conclusion of the 9/11 Commission that 'no credible evidence' for such a link existed, and the CIA's determination that [Saddam] Hussein 'did not have a relationship' with Al Qaeda—the Vice President continued to insist that the relationship had been confirmed" and pressed for war.

It's worth noting that in the five years Dick Cheney had worked for a big defense contractor, Halliburton, before becoming Vice President of the United States, he received $12.5 million in salary, and when he left the company, he got a $34 million dollar exit package from Halliburton—a fortune for a man with no previous experience in running a multinational corporation. Plus, Halliburton paid Cheney an extra $1 million during the time he served as Vice President.

After Cheney convinced George W. Bush to launch a pre-emptive and illegal attack on Iraq, Halliburton was chosen as the go-to contractor for the war, for which the company earned $39 billion. Senator John Kerry, who later served as Secretary of State under Barack Obama, charged that "Dick Cheney's old company, Halliburton, has profited from the mess in Iraq at the expense of American troops and taxpayers." No weapons of mass destruction were found in Iraq, despite Cheney's claims to the contrary. It's deplorably ironic that Chickenhawk Cheney got five deferments from the draft to avoid military service, yet was the Bush administration's chief warmonger.

Writing in *The Atlantic*, Conor Friedersdorf said "Dick Cheney was a self-aggrandizing criminal who used his knowledge as a Washington insider to subvert both informed public debate about matters of war and peace and to manipulate Presidential decisionmaking, sometimes in ways that angered even George W. Bush."

While presiding over Halliburton, writes Friedersdorf, "he did business with corrupt Arab autocrats, including some in countries that were enemies of the United States. Upon returning to government, he advanced a theory of the executive that is at odds with the intentions of the Founders, successfully encouraged the federal government to illegally spy on innocent Americans, passed on to the public false information about weapons of mass destruction in Iraq, and became directly complicit in a regime of torture for which he should be in jail.

"When Vice President Dick Cheney left office, his approval rating stood at a staggeringly low 13 percent. Few political figures in history have been so reviled," concludes *The Atlantic*.

Given all this, is it any surprise that Dick Cheney would advocate for environmental policies that would profit his benefactors?

A Bush/Cheney sinister plot to plunder our public lands

Robert F. Kennedy, Jr., who has been a dedicated environmental lawyer for more than 32 years, pointed out that "President Bush appointed as head of the Forest Service a timber industry lobbyist, Mark Rey—probably the most rapacious in history. He put in charge of public lands a mining industry lobbyist, Steven Griles, who believes that public lands are unconstitutional. He put in charge of the air division of the EPA, Jeffrey Holmstead, a utility lobbyist who has represented nothing but the worst air polluters in America. As head of Superfund, a woman whose last job was teaching corporate polluters how to evade Superfund. The second in command of EPA [was] a Monsanto lobbyist."

Kennedy recalls that to advise him on environmental policy, Bush "put a lobbyist of the American Petroleum Institute whose only job was to read all of the science from all the different federal agencies to make sure they didn't say anything critical, to excise any critical statements about the oil industry."

In the first three years of his Presidency, Mr. Bush initiated more than 200 major rollbacks of America's environmental laws, weakening the protection of our country's air, water, public lands and wildlife. Under the guidance of Republican pollster Frank Luntz, the Bush White House disingenuously hid its anti-environmental program behind deceptive, Orwellian rhetoric, telegenic spokespeople, secrecy and the intimidation of scientists and park rangers, who were afraid of being fired and losing their pensions if they spoke out.

With little public debate, the Bush Administration approved a plan to drill 66,000 coalbed methane gas wells in the Powder River Basin of Wyoming and Montana—a massive project that resulted in 26,000 miles of new roads, 48,000 miles of new pipelines, and discharges of two trillion gallons of contaminated water, disfiguring for years the rolling hills of that landscape.

Despite all his pious political rhetoric and entreaties to the so-called "religious right," Mr. Bush defied Biblical injunctions about good stewardship of the land, and ignored a broad national consensus that the quality of our lives in inextricably linked to the preservation of our natural habitat.

He pursued a furtive campaign to place the interests of corporate America ahead of the public good, and in so doing irreversibly squandered some of our most precious natural resources.

His actions drew over 50,000 public comments and concerns from former Park Service employees, lawmakers of both parties, many outdoor businesses, and the National Council of Churches.

Unethically, the Bush White House used stealth tactics, such as encouraging states and private groups to file lawsuits against the federal government, and then agreeing to negotiated settlements that bypassed environmental laws without any oversight from Congress or

The Tetons and Snake River, 1942: The iconic photographs of Ansel Adams captured our nation's pristine wilderness, a treasure Americans want preserved.

input from the general public.

As environmental writer Chip Ward observed, "Everywhere you look, the Bush war on the environment defies public opinion. Under cover of the weekend in March 2003 when he invaded Iraq, Bush's Secretary of the Interior Gale Norton signed a Memorandum of Understanding with Utah's then-Governor Mike Leavitt (who went on to head the EPA and was later Secretary of Health). It allowed the state to absurdly claim that thousands of dirt tracks and paths through public lands were actually 'highways' under an obscure law designed in the 19th century to allow prospectors access to mining claims."

The out-of-court settlement of a lawsuit, signed that same weekend by Norton and Leavitt, stripped millions of acres of public lands throughout the West of safeguards that helped maintain their pristine character. In Utah alone, six million acres of land that meet all the criteria for wilderness designation and protection can no longer be managed that way.

Private lands were also affected. In the west, ranchers have had their gates bulldozed away by oil drillers for corporations that own surface mining rights and now feel free to take their heavy equipment into privately operated ranches without permission or notification.

Established by the Wilderness Act of 1964, the very concept of "wilderness," the most popular and important conservation tool ever created, has now been stripped of its meaning and power.

The actions of Norton and Leavitt proved typical of Bush-era strategies meant to implement otherwise unpopular decisions that could not stand up to public scrutiny or involvement—or survive legal challenges, even in courts packed with Bush-friendly judges. We can guess who came up with this strategy.

Privatization Run Amok

The credo of the George W. Bush White House was to "outsource" many of the Park Service's critical functions, including biological science and archaeological survey and assessment activities, which replaced Park Service workers with low-bid private contractors. But outsourcing fails to place any value on the expertise and institutional knowledge of Park Service professionals, such as archaeologists who are responsible for preserving Civil War battlefields, prehistoric ruins and artifacts, dinosaur bones, fossils and other relics of American history. As a result, shifting worker duties to private industry can actually increase costs over retaining Park employees because of the loss in productivity and training time, not to mention the loss of educational benefits to visitors. Privatization also further opens National Park management to private influence, rather than retaining direct government oversight and at least the veneer of objectivity when weighing the public interest.

"The Bush administration's outsourcing venture [was] a bust, costing taxpayers millions of dollars and hurting employee morale to determine that, by and large, the government workforce does a fine job at a fraction of the cost of private corporations," said Greg Wetstone, of the NRDC. "In the case of the Forest Service, in particular, it's not a good idea to eliminate civil servants whose job it is to serve the public and tell the truth with a P.R. firm paid to spin the Administration's story."

Under Bush/Cheney, the coal industry was given leave to scalp the tops off the Appalachian Mountains—historic landscapes where Daniel Boone and Davy Crockett roamed, and that are the source of our values and culture. Giant machines called drag lines that are 22-stories high and cost half a billion dollars are extracting coal and scarring the terrain. The residue is being dumped into valleys, obstructing

Mountaintop removal mining devastates the landscape, turning areas that should be lush with forests and wildlife into barren moonscapes.

streams, killing aquatic life, and occasionally triggering mudslides and floods.

Mountaintop removal coal mining, often described as "strip mining on steroids," is an extremely destructive form of mining that continues to devastate Appalachia.

Over 2,000 miles of streams and headwaters that provide drinking water for millions of Americans have been permanently buried and destroyed. An area the size of Delaware has been flattened. Local coal field communities routinely face devastating floods and adverse health effects. Natural habitats in some our country's oldest forests are laid to waste.

As Earth Justice explains the process, "Coal companies first raze an entire mountainside, ripping trees from the ground and clearing brush with huge tractors. This debris is then set ablaze as deep holes are dug for explosives.

"An explosive is poured into these holes and mountaintops are literally blown apart. Huge machines called draglines—some the size of an entire city block, able to scoop up to 100 tons in a single load—push rock and dirt into nearby streams and valleys, forever burying waterways.

"Coal companies use explosives to blast as much as 800 to 1,000 feet off the tops of mountains in order to reach thin coal seams buried deep below."

In 2009, after President Obama took office, there was a great deal of optimism among Appalachian citizens and those concerned about mountaintop removal. New agency heads and White House spokespersons parroted the talking point that "the administration will do what the science calls for." In Appalachia, the science calls for an end to mountaintop removal coal mining.

Obama's people did far more to curtail the destructive mining practice than either the Clinton or Bush administrations. While that's a low bar, they did make some significant changes, and there was less mountaintop removal mining during Obama's term in office than there was between 2002 and 2008 when Bush was in the White House. Part of that is due to market forces, and part is attributable to the actions of the Obama administration. These actions, however, have not been enough. Instead of refusing all permits associated with mountaintop removal mining, Obama continued to issue permits for mines and valley fills. While Obama issued fewer permits than his predecessors, he issued permits nonetheless. Modest steps are not good enough.

Nowhere are the results of Bush's assault on our natural resources more manifest than in our National Parks.

Where things stand now

Bush/Cheney can be blamed for the new rules that emphasize recreation over preservation that have resulted in off-road vehicles harming archeological sites at the Grand Canyon; tearing up hiking and horseback trails in Olympic National Park, crushing animal burrows in the Canyonlands National Park, and facilitating fossil poaching at Badlands National Park in South Dakota. Park managers at the Appalachian Trail say that off-road vehicles are still the Trail's "most pernicious" problem, even after Barack Obama took steps to try to ameliorate some of the difficulties. But there are other issues in 2019:

•At Yellowstone, 150 miles of roads have not been repaired in years, and many of the park's several hundred buildings are in poor condition. The park's 273-foot Lewis River Bridge, more than half a century old, has reached the point of requiring either "extensive rehabilitation" or complete replacement.

Routes that carry visitors to Yellowstone's geothermal sites, waterfalls, and wilderness require at least $1 billion worth of reconstruction. The park contains 254 miles of main roadway, but at current funding levels it will take 75 years to finish the work, said Superintendent Dan Wenk.

• Yosemite National Park needs more than $582 million for backlogged projects, including trail and campground maintenance.

• Hikers cannot reach backcountry cabins at Mount Rainier National Park in Washington State because bridges and trails need repair.

• The 469-mile Blue Ridge Parkway needs $350 million worth of work, but is allocated only seven to ten million dollars a year in federal highway funds. The Appalachian roadway, and its more than 150 bridges and 25 tunnels, suffer erosion from the sides and underneath as broken-down drainage structures fail to divert stormwater away.

• Four out of ten historic buildings at Gettysburg's hallowed battlefields in Pennsylvania and the neighboring Eisenhower National Historic Site are in poor or serious condition.

• Large sections of a historic lighthouse and Fort Jefferson at Dry Tortugas National Park in South Florida are unsafe.

• Ancient stone structures are collapsing at Chaco Culture National Historical Park in New Mexico.

• Researchers believe anglers have introduced non-native earthworms into Voyageurs National Park in Minnesota. The earthworms change the soil, which changes the trees, which affects water that flows into lakes.

• In Florida, the fast-draining Everglades are affected by an average of 900 new Florida residents a day who create a daily new demand for 200,000 gallons of water. Visitors driving into Everglades National Park through vast carpets of grasses and cypress trees, have to swerve around road bumps and eroded surface.

• At Sequoia National Park, writes Rita Beamish, "Cars crawl up the steep, single-lane road, winding between eroded edges and over pavement cracks, with cliff drop-offs and stunning vistas emerging around bend after bend. Drivers need to focus, and that means resisting the scenic distractions of burbling streams below and Sequoia tree giants stretching above."

Ms. Beamish is referring to "Mineral King Road, built to reach mining claims in the late 1800s, a 15.2-mile gateway to a Sierra Nevada valley so appealing that none other than Walt Disney once sought to put a massive ski resort here."

If you drive into the valley, it's like entering a time warp, back to the turn of the last century. Much of the road remains as constructed in 1870. Woody Smeck, superintendent of Sequoia and Kings Canyon

National Parks said, "That's a bumpy road that needs a lot of work. I worry about it because it's narrow and winding from a traffic safety point of view."

• Shamefully, even the visitor center at the *USS Arizona* Memorial at Pearl Harbor in Hawaii was being allowed to sink. The Memorial remains in a frustrating, year-long closure for dock repairs that started in May 2018. The memorial itself will remain closed likely until just before December 7, 2019.

The National Park service has failed to give regular updates on the progress of repairs. The *USS Arizona* Memorial is the most iconic part of Pearl Harbor, which is to this day the number one visitor destination in Hawaii. What happened at Pearl Harbor on December 7, 1941, resulted in America's entry into World War II on the following day.

Crime Overtaking the Parks

Drug smugglers and illegal immigrants have invaded our National Parks. Statistics show that park rangers are the most assaulted of all federal law enforcement officers.

Some poetic language on the National Park Service website describes an idyllic location as follows: "There is a place in Far West Texas where night skies are dark as coal and rivers carve temple-like canyons in ancient limestone. Here, at the end of the road, hundreds of bird species take refuge in a solitary mountain range surrounded by weather-beaten desert. Tenacious cactus bloom in sublime southwestern sun, and diversity of species is the best in the country. This magical place is Big Bend..."

Stretching from east of El Paso nearly to Del Rio in the south, and northward through the Panhandle and all of Oklahoma, the Big Bend National Park is larger than California, encompassing more than 165,000 square miles.

But there are only about 500 law enforcement agents to patrol the sector, the vast majority in the mountainous and rugged Big Bend, which straddles a traditional travel corridor between northern Mexico and the United States and has no border walls.

In contrast, more than 3,000 agents work in the heavily trafficked Rio Grande Valley Sector, and almost 4,000 work in the Tucson Sector.

In the past, during Prohibition, the smugglers' drug of choice was alcohol and confrontations occurred fairly routinely with bootleggers, who were arrested.

Historian Cecilia Thompson says that the diary of one border patrol agent in the 1920s noted that "he encountered German, English, Canadian, Syrian, Russian, Jewish, Turkish, Greek, Hungarian and Austrian aliens, as well as Mexican aliens, frequently traveling on foot across the barren wastes of the Big Bend." So illegal border crossings are nothing new.

Today, most agents in Big Bend must wear bulletproof vests while on duty and carry both a handgun and semi-automatic weapon. Agents in some areas of Big Bend still use horses to patrol in rough terrain. Helicopters are deployed for daytime searches, and at night, agents use infrared devices to peer through the darkness and spot smugglers. One big improvement is, agents now have the technical means to make instantaneous identifications of criminal suspects in the field if the criminals have fingerprints on file from previous arrests.

• In Texas, at the Padre Island National Seashore, drug smuggling, illegal immigrants, the poaching of endangered turtles and their eggs, and unlawful commercial fishing pose a threat to the park's resources, its visitors and the rangers themselves. The lack of timely backup for officers in trouble, and sometimes erratic radio communications, are issues at this location.

Another complication is that oil and natural gas drilling is allowed within the park. Congress has not approved the purchase of the mineral rights within the park even though the boundaries were sur-

A geological marvel: The Balanced Rock, Big Bend National Park.

veyed as early as 1957. This has caused some controversy because the land is a protected seascape. Yet heavy equipment is used within the park and is transported across beaches that are nesting areas for not only sea turtles but many birds, including the piping plover, least tern, and other animals that may nest within the park.

• Organ Pipe Cactus National Monument on the southern Arizona border was listed as the most dangerous National Park. One park ranger was shot and killed while chasing a drug cartel hit squad. So starting in 2003, the park was closed for 11 years, only re-opening in 2014. This was due to numerous incidents involving international drug trafficking, the inflow of illegal immigrants, and a workforce that is understaffed to safely manage the problem. The park also reported that criminals had created miles of illegal roads in the park. As many as 1,000 aliens and drug smugglers pour into Arizona's Organ Pipe daily, diverting 75 percent of rangers' time to the problem.

On the U.S. side of the border, promotion for Organ Pipe extols "Immerse yourself in a photographer's paradise! Explore the abundance of plants and wildlife unique to the Sonoran desert. Guided walks through the park, as well as hiking trails, camping and picnic facilities, are available. Drive the scenic 21-mile Ajo Mountain loop...star-studded night skies wash away the modern world."

However, on the other side of the border, in Mexican towns adjacent to the park, are pamphlets that offer illegal border aliens tips in Spanish:

"Use the north star and the movement of the moon to guide you towards the north during the night. Carry one gallon of water in each hand and six litres in the backpack. You can drink cactus fruit but the skin has nearly invisible spines. Peel carefully. If you have no water, drinking urine can sustain you for a while. Don't do it repeatedly because it will become toxic."

The crush of human traffic has left a trail of ravaged vegetation and human excrement. "Some areas, the smell of the human waste just hits you," said one park ranger. "It's overwhelming right now and it's not safe for our staff to go out and start a cleanup."

Despite the dangers of the park, there are "No more armed guards," says Sue Walter, the monument's chief of interpretation. "[The border] has surveillance towers, vehicle barriers, pedestrian fences. We're educating visitors and they can make their own decisions about whether they feel comfortable [going into the backcountry]."

Organ Pipe, 94 percent of which is designated wilderness, once was traveled mostly by the endangered Sonoran pronghorn and Sonoran desert tortoise. It began drawing both human and drug traffickers in the 1990s, after border security crackdowns in urban areas sent crossers to remote rural areas. As Yahoo News reported, rangers routinely found themselves in high-speed chases, and they seized thousands of pounds of marijuana and other drugs each year—17,000 pounds of pot in 2005, 100,000 pounds in 2014. Not surprisingly, the monument's reputation tended to scare away visitors. Now, about 200,000 people visit the park each year.

Harm has also been done to the area in the name of national security. The feds can waive environmental laws that represent 40 years of work and achievement, and take away the rights of citizens to protest it.

Vehicle tires leave lasting scars wherever anyone cuts across this desert, breaking soil crusts that were formed over hundreds, thousands, or even a million years. "The soil turns to talc when you drive over it," says Todd Esque, an ecologist who tested soil impacts in northwest Organ Pipe Cactus Monument. Driving an ATV or truck just once across that soil has more impact than 50 hikers, creating localized "Dust Bowl" conditions. And once a single driver cuts across the desert, a road starts, it doesn't end. More and more drivers use it.

•The Lake Mead National Recreation Area in Nevada and Arizona has ongoing gang activity, and vast areas of backcountry have only cursory patrols. There are fewer park rangers working this year than last, and some of those that remain have been rotated out to provide security at the dams.

Arizona depends on Lake Mead to provide about 40 percent of its water supply each year. But the lake is rapidly careening toward a shortage, which could occur as early as 2020, forcing Arizona to take a significant water allocation cut.

The problem is that Arizona, California, Nevada and Mexico have the rights to millions of acre-feet more water than melting snow and rainful replenishes each year due to climate change. Nearby cities could be forced to implement their drought mitigation plans to turn off fountains, stop overseeding parks and golf courses, and take certain other measures to save water.

Pollution is overwhelming

Pollution that has drifted scores of miles into parks is affecting visitors, plant life and wildlife. Last year, the air breathed by park visitors exceeded eight-hour safe levels of ozone 150 times in 13 parks, from California to Virginia. Overall, air at one-third of parks moni-

Hikers along a ridge in Maine's Acadia National Park in the NE part of the U.S. sometimes get a clear view but the air is among the most polluted in the nation.

tored by the Park Service continues to worsen even as the government puts in place pollution controls aimed at clearing the air by 2064—an extraordinarily long time to wait!

The National Park website for Air Quality at Sequoia National Park says the park is one of the worst national parks for air pollution in the country. The reason is that the park is "downwind of many air pollution sources, including agriculture, industry, major highways, and urban pollutants from as far away as the San Francisco Bay Area," according to the website.

A new study shows that in all but two years since 1996, Sequoia National Park ozone exceedance days have surpassed that of Los Angeles, the metropolitan area with the highest ozone concentration.

"You can see this in the parks' Twitter feed," said Ivan Rudik, co-author of the study and assistant professor of applied economics at Cornell University. He noted that "pretty much every day in the summer there is an air quality warning where it is unhealthy for sensitive groups or for all people."

According to the Grand Canyon National Park air profile page, it "is downwind of air pollution from coal-fired power plants in the Four Corners region, nearby mining, and urban and industrial pollutants from Mexico and California."

On the other side of the nation, Great Smoky Mountains National Park in Tennessee and North Carolina, the most frequently visited park, has air quality similar to that of Los Angeles.

Many others, including Shenandoah in Virginia, Mammoth Cave in Kentucky, Sequoia and Kings Canyon in California and Acadia in Maine also suffer reduced views and damage to natural resources, mostly from pollutants from coal-fired power plants.

Acadia National Park's page says the pollution drifts in from large urban and industrial areas in the states to the south and west. This polluted air gets trapped in the park's steep slopes and high peaks. The park's air page says after 30 years of air quality monitoring, it has shown the park "receives some of the highest levels of pollution in the northeastern U.S."

Besides making the air unsafe to breathe, pollution has diminished the average daytime visibility from 90 miles to less than 25 miles at Eastern parks, and in the West from 140 miles to between 35 miles and 90 miles, the Environmental Protection Agency said.

Climate change threatens our parks, too

Even global warming is harming our parks, putting 12 of the most famous U.S. National Parks at risk. This conjures up visions of Glacier National Park without glaciers and Yellowstone Park without grizzly bears. All 12 parks are located in the American West, where temperatures have risen twice as fast as in the rest of the United States over the last 50 years, said Theo Spencer of the Natural Resources Defense Council (NRDC).

"Rising temperatures, drought, wildfires and diminished snowfalls endanger wildlife and threaten hiking, fishing and other recreational activities" in the parks, Spencer said.

Commercial development gobbles up wilderness areas

Since 1916, the National Park Service has had a mandate "to protect and preserve unimpaired the resources and values of the National Park system." This means clean air, wilderness protection, unspoiled vistas, and wildlife conservation.

But as the Associated Press reports, "in an age when many Americans expect homelike amenities while they're enjoying nature, when they prefer the option of sightseeing from low-flying aircraft or snowmobiles, and demand constant cell phone service, this is not an easy charter to fulfill. Especially since the U.S. population has grown more than 200% since 1916, and budget cuts have left many parks strapped for funds."

Parks are being opened up to more commercial activities, livestock grazing, and motorized recreation. Housing developments and strip malls are springing up just outside of parks.

There already are dozens of cell phone towers inside the 390 national park units, including within sight of Yellowstone's Old Faithful geyser. Tourist-filled helicopters and airplanes buzz around the Grand Canyon. The sounds of snowmobiles and personal watercraft still break the pristine silence in some parks.

In 2007, George W. Bush changed park policy to allow as many as 720 snowmobiles a day into Yellowstone, and 140 in neighboring Grand Teton National Park.

Vacation homes now dot the shores lining Acadia and the mountains that border the Great Smoky Mountains National Park. Subdivisions have sprouted up around hallowed Civil War sites such as Manassas Battlefield Park in Virginia.

Gettysburg National Military Park sees more than a million vis-

Unbelievable sight: While President, John Kennedy drove himself, his wife Jackie and friends on a tour of the Gettysburg battlefield where Lincoln spoke and 60,000 men died. Now this National Park is threatened by nearby commercial development.

itors each year, most of them tourists from within the U.S. To keep this onslaught of people entertained when they're not engaged in solemn commemoration of the epic battle where 60,000 soldiers died in 1863, and Abraham Lincoln delivered his most famous speech, the area near Gettysburg is full of historic sites, museums, farms and other activities. But though corn mazes and petting zoos are in keeping with local character, many argued that gambling was not. Yet a casino has repeatedly been proposed—and rejected—within cannon range of the Gettysburg battlefield.

Several hundred new homes have been constructed along the scenic New River Gorge National River in West Virginia.

Even the parks' famed views of starry skies are in jeopardy. Nighttime lights, beaming from cities and towns 200 miles away from parks such as Mount Rainier in Washington state and Yosemite in California, reduce star visibility and can affect nocturnal wildlife.

In urban regions like the Santa Monica Mountains National Recreation Area in California, visitors can only see a few hundred stars instead of the 8,000 that would be visible in pristine conditions.

"If there's no place that is clear and clean, if there's no place that is dark and starry, where does that leave us?" asks Chad Moore, program manager for the National Park Service's Night Sky Team. "If we can't protect the best parts of America in National Parks, then we're certainly not going to be able to protect them anywhere else."

A Backlash to Radical Policies

The GOP policies have met increasing opposition from within their own ranks. Republicans for Environmental Protection, (REP) was founded to counter the party's ideological departure from its conservation heritage of Theodore Roosevelt and restore its traditional ethic of environmental stewardship.

Rev. Tri Robinson, a pastor at a dynamic conservative church in Boise, Idaho, attracted national attention by preaching to his congregation the Biblical injunction to treat the earth with respect.

In the Appalachian Mountains of West Virginia, local residents brought their faith to bear in their effort to stop widespread pollution and environmental damage. In a state where three million pounds of explosives are used each day to strip the mountains of their coal, some evangelicals are relying on scripture to battle Big Coal.

"I first want to apologize as a Christian for the unfaithfulness of the churches and Christians who have oftentimes—too often—been complicit in the destruction that we see upon the land," says Allen

Johnson, who co-founded the advocacy group Christians for the Mountains. "In the Book of Revelation, there's a scripture that says that God will destroy those who destroy the Earth. We're breaking a covenant with God."

Barack Obama brings relief

In sharp contrast to his predecessor, President Barack Obama made protecting the environment and combating climate change one of the cornerstones of his Presidency. He said many times that he "believes that no challenge poses a greater threat to our children, our planet and future generations than climate change—and that no other country on earth is better equipped to lead the world towards a solution."

Knowing that green initiatives are historically unlikely to win bipartisan support, he once pledged that "if Congress won't act soon to protect future generations, I will." And he made good on that promise. Although Obama fell short of what many environmentalists had hoped he would accomplish, the President did make some remarkable progress in protecting our vanishing wilderness:

More than any other President, Obama used his authority under the Antiquities Act of 1906: This allows U.S. Presidents to make a Presidential proclamation to create national monuments from public lands to protect significant natural, cultural or scientific landmarks. Obama did this 23 times. For example, in New York City, he designated Christopher Park and renamed it the Stonewall National Monument to pay homage to the Stonewall riots that happened nearby in 1968, launching the modern Gay Rights Movement. And in New Mexico, he designated the Rio Grande del Norte National Monument, which according to the Bureau of Land Management (BLM) "is comprised of rugged, wide open plains at an average elevation of 7,000 feet, dotted by volcanic cones, and cut by steep canyons with rivers tucked away in their depths."

Signed the Omnibus Public Land Management Act of 2009: The White House called this largely bipartisan bill the "most extensive expansion of land and water conservation in more than a generation, designating more than two million acres of federal wilderness, thousands of miles of trails and protecting more than 1,000 miles of rivers."

When land is designated as federal wilderness, it receives the highest form of protection of any federal wildland and becomes part of the National Wilderness Preservation System, according to the Wilderness Society. Nine states—California, Colorado, Idaho, Michigan, New Mexico, Oregon, Utah, Virginia and West Virginia—were included in the bill.

Established America's Great Outdoors Initiative: In 2010, President Obama launched this initiative to develop a "21st century conservation and recreation strategy." And by November 2011, the Department of the Interior had released a listing of projects in all 50 U.S. states, including the Rio Salado River Pathways Program in Arizona and the Yampa River Basin Project in Colorado. According to the Bureau of Land Management (BLM), some of the key goals were:
• Enhance recreational access and opportunities.
• Raise awareness of the value and benefits of America's great outdoors.
• Engage young people in conservation and the great outdoors.
• Establish great urban parks and community green spaces.
• Conserve rural working farms, ranches and forests through partnerships and incentives.
• Conserve and restore our federal lands and waters.
• Protect and renew rivers and other waters.

Rejected Keystone XL oil pipeline: On Nov. 6, 2015, Obama rejected construction of the 1,179-mile pipeline from Canada to the Texas Coast that became a symbol of the debate over his climate change policies. After a seven-year review, the President nixed the TransCanada proposal which Republicans promised would create thousands of new jobs, and said: "America is now a global leader when it comes to taking serious action to fight climate change. Frankly, approving this project would have undercut that global leadership."

The $8 billion project would transport heavy crude from Alberta, Canada to Steel City, Nebraska, where it would link up with existing lines to transport oil supplies to Gulf Coast refineries. The pipeline would carry oil sands, a type of crude that former Secretary of State John Kerry called "a particularly dirty source of fuel."

In one of his first acts as president, Donald Trump signed an executive order to advance Keystone XL and another controversial pipeline, Dakota Access, which started service over a year ago.

But in November 2018, a federal judge in Montana, Brian Morris, ruled that the Trump administration failed to conduct the necessary environmental reviews when it approved Keystone XL in 2017. In February 2019, Morris dealt the Keystone XL pipeline another setback, effectively blocking the oil pipeline company, TransCanada's latest effort to begin work on the project. Also, the Nebraska Supreme Court is considering a lawsuit over the pipeline route approved by the state's Public Service Commission brought by landowners.

CNBC reports that "Keystone XL has become a major flash point in the growing war between environmentalists and the oil industry over expanding pipeline infrastructure. The line has now been tangled up in political battles for about a decade."

"It's been two years since the Trump administration tried to revive this pipeline from the dead, but Keystone XL is still far from being built," Jackie Prange, senior attorney at Natural Resource Defense Council said in a statement. "Today's decision is one more victory for the rule of law over this reckless and risky project."

Expanded the California Coastal National Monument: In March 2014, Obama established the first shoreline addition to the monument, which consists of 20,000 rocks, islands, exposed reefs and pinnacles along 1,100 miles of Northern California's coast. The addition is a swath of rocky shores, river corridors, meadows and wetlands known as Point Arena-Stornetta Public Lands. "In my State of the Union address, I said that I would use my authority to protect more of our pristine federal lands for future generations," he said in a statement. "Our country is blessed with some of the most beautiful landscapes in the world. It's up to us to protect them, so our children's children can experience them, too."

Every Kid in a Park initiative: Extended beyond his Presidency, the program aims to get more children playing and learning outdoors by providing 4th grade students and their families free admission to all National Parks and other federal lands and waters.

Started a plan to save bees and pollinators: In 2014, Obama created a "Federal Strategy to Promote the Health of Honey Bees and Other Pollinators" and established the Pollinator Health Task Force, which was charged with developing a National Pollinator Health Strategy, which it released a year later. The goals of the strategy, according to the EPA, are:

• Restore honeybee colony health to sustainable levels by 2025.

• Increase Eastern monarch butterfly populations to 225 million butterflies by 2020.

• Restore or enhance seven million acres of land for pollinators over the next five years.

Concerns about the future

The Donald Trump administration has made it abundantly clear that one goal is to systematically sell out America's public lands to the fossil fuel industry—much more so than Obama allowed.

In collaboration with Congressional Republicans, Trump has launched a coordinated and calculated attack on the fundamental laws and policies that guide the sustainable, multiple-use management of these national assets. The cumulative effect of this could set back public-lands management for decades.

The onslaught began soon after the 2016 elections, when Republicans used the Congressional Review Act to rescind sensible Obama-era planning rules for public-land management. They included actions aimed at curbing the methane emissions that contribute to climate change, as well as efforts to protect waterways impacted by coal mining. The Administration has reversed decisions to prevent construction of the Pebble Mine in Alaska and to curb mining that threatens the Boundary Waters Canoe Area in Minnesota. What's more, the Republican members of the Montana and Oregon Congressional delegation are advancing legislation to eliminate wilderness study areas in their states.

Jim Lyons, a lecturer at the Yale School of Forestry and Environmental Studies and senior fellow at the Center for American Progress, complains that "More bad policy emerged when the Bureau of Land Management and the U.S. Fish and Wildlife Service reversed existing mitigation policies. Now, land users who damage the public's lands in developing mineral or energy resources will no longer be required to repair the damage they cause."

A major new sell-off of public lands

Recently, hundreds of corporate representatives sat down at their computers, logged into something called *Energynet*, and bid, *eBay* style, for more than 300,000 acres of federal land spread across five Western states. They paid as little as $2 per acre for control of parcels in southeastern Utah's canyon country, Wyoming sage grouse territory and Native American ancestral homelands in New Mexico.

As Jonathan Thompson reports in *Perspective*, "Even as public land advocates scoff at the idea of broad transfers of federal land to states and private interests, this less-noticed conveyance continues unabated. It is a slightly less egregious version of the land transfers that state supremacists, Sagebrush Rebels and privatization advocates have pushed for since the 1970s.

"It's called oil and gas leasing, conducted under the Mineral Leasing Act of 1920. With President Donald Trump touting in his State of the Union speech...that Republicans have 'ended the war on American energy,' you can expect leasing to ramp up in years to come."

Meanwhile, as they say, the devil is in the details. George Nickas reports that "when the likes of anti-public lands legislators Senator Lisa Murkowski (R-AK) and Representative Rob Bishop (R-UT) stamp their approval on a massive 698-page public lands omnibus bill, we'd best dig deep. So, why isn't that happening? A bipartisan chorus has applauded the 'Natural Resources Management Act,' a bill written in the last Congress—the most anti-public lands Congress in memory—and about to be rubber-stamped by the new one. It is being hailed as one of the biggest conservation achievements in decades, but it is full of harmful provisions that would never see the light of day were they not tucked quietly into the omnibus.

"Take the relatively innocuous sounding 'wildlife management in National Parks' provision. It should be called 'Opening National Parks to Hunting,' because that's what it does. It allows the Secretary of Interior...to open the Parks to 'volunteer' hunters whenever the Secretary deems a wildlife population needs culling. Former Sec. Zinke had already made such a declaration for predators in national preserves in Alaska, where state officials are pushing to eliminate wolves, grizzly bears, and anything else that eats hunters' 'game.' There's little reason to believe [Zinke's successor] and his ilk won't do the same elsewhere. In the states surrounding Yellowstone National Park there's a constant cry from state officials to cull the bison and elk herds, and to limit the number of wolves and grizzly bears that dare wander beyond the Park borders. [Trophy hunters] in groups like Safari Club International and the NRA have always chafed at the ban on hunting in National Parks, and the public lands bill is their key to finally opening the lock. And it's not limited to just Yellowstone. Bison in the Grand Canyon, elk in Rocky Mountain and wildlife in other parks could become targets with passage of the bill."

George Nickas warns there's more. "The so-called 'sportsmen's' provision elevates hunting, angling, and recreational shooting as a priority in public lands management. A major gas pipeline will run through Denali National Park. Other provisions bring many new problems for our National Wilderness Preservation System. What did you expect, given the previous Congress wrote the bill?

"To be sure, the bill contains positive provisions, but it should have undergone the scrutiny of committee hearings, public hearings, and proper oversight." Nickas says "The U.S. House of Representatives should do just those things before the bill becomes law, or if the ship is too big to steer at this point, perhaps we should hail an iceberg.

"They say it's a done deal, and it probably is. But if you want to contact your Member of Congress and express your concerns, you can reach their offices at 202-224-3121."

Trump's government shutdown also damaged our parks

The shutdown of the federal government by President Donald Trump in one of his more autocratic and tantrum-like moves to try to force Congress to fund his border wall (that even members of his own party rejected), resulted in not only widespread harm to the country's economy but its environment as well.

Besides furloughing hundreds of thousands of federal government employees and private contractors, who were left without income, Trump's irresponsible actions also left many parks without most of the rangers and others who staff campgrounds and otherwise keep parks running. Trash was piling up and overflowing, threatening the safety of visitors and harming wildlife. Vehicle accidents were occurring due to the lack of staff to sand the icy roads during wintry conditions.

USA Today reported that "Human feces, overflowing garbage, illegal off-roading and other damaging behavior in fragile areas were beginning to overwhelm some of the West's iconic national parks."

"It's a free-for-all," Dakota Snider, 24, who lives and works in

Joshua Tree National Park was vandalized during the government shutdown.

Yosemite Valley, said. "It's so heartbreaking. There is more trash and human waste and disregard for the rules than I've seen in my four years living here."

Unlike previous government shutdowns, which also closed recreation areas, Trump left the National Parks open. It was as if the President was deliberately trying to do more harm to our nation's natural resources.

USA Today quoted John Garder, senior budget director of the nonprofit National Parks Conservation Association, as proclaiming "It's really a nightmare scenario. We're afraid that we're going to start seeing significant damage to the natural resources in parks and potentially to historic and other cultural artifacts," Garder said. "We're concerned there'll be impacts to visitors' safety."

At Joshua Tree National Park, some visitors strung Christmas lights in the twisting trees, many of which are hundreds of years old. Most visitors were being respectful of the desert wilderness and park facilities, Joshua Tree's superintendent, David Smith, said in a statement.

But some were seizing on the shortage of park staffers to off-road illegally and otherwise damage the park, as well as relieving themselves in the open.

The former superintendent of Joshua Tree National Park, Curt Sauer, said at a rally to save the park that it could take hundreds of years to recover from damage caused by visitors during the longest-ever government shutdown.

"What's happened to our park in the last 34 days is irreparable for the next 200 to 300 years," he declared. Volunteers hauled out trash and cleaned toilets—but in a park covering 1,235 square miles, it wasn't enough to stop damage from visitors to the park who did not heed warnings to take extra care.

Delicate desert ecosystems in Death Valley have been damaged by off-roaders, another dismaying impact of the U.S. government shutdown on national parks.

"People come here to this pristine desert landscape," said Laura Cunningham, who heads Western Watersheds Project, a not-for-profit conservation organization. She and her husband, a retired Death Valley park ranger, live close to the park and headed out to the desert to assess new damage. "There are so few places where we have a beautiful natural vista. And now people are off-roading on it."

Cunningham and her husband took photos of what they saw: tire marks etched into delicate playas and plains that can take centuries

Tire tracks in forbidden areas defaced Death Valley.

to recover. Photo evidence of this destruction appears above.

She said there was evidence that several vehicles had careened through the colorful badlands and through the hills where animals— foxes, kangaroo rats and an assortment of lizards—burrow underground. "They can be killed when you just drive over the ground," she said.

"We are worried about the more remote places where people are driving off into this very sensitive playa."

The park service calls Death Valley an "outdoor natural history museum" with "examples of most of the earth's geological eras and the forces that expose them."

National Geographic reported that parks lost roughly $400,000 a day in entrance fees during Trump's government shutdown, and now more funds will be needed to make repairs.

"Some [efforts] will take weeks or months. Some will last generations. Some may not be able to be fixed," the former National Park Service director Jonathan Jarvis told the publication in early January 2019.

The night sky is no longer dark in some areas

Chief among these is the Theodore Roosevelt National Park, a lesser-known park nestled in the badlands of North Dakota. It has traditionally been known for its dark skies that attract stargazers. But now,

Mount Rushmore National Memorial in the Black Hills of South Dakota. At one point proponents wanted to add Ronald Reagan to the four Presidents—Washington, Jefferson, Roosevelt and Lincoln—but that idea met with a hostile public reaction.

thanks to fracking activity right outside the park, "It looks like Blade Runner out there, with gas flares going off and trucks rolling by," says James Nations, who leads the NPCA's Center for Park Research and edited the report.

The gas flares and lights around drilling rigs have polluted the night sky across the 70,000-acre park, which is comprised of three parcels spread around the massive, 128 million-acre Bakken formation. That says nothing about concerns over what the chemical-heavy fracking fluid might be doing to water sources and what the noise and road-building from drilling operations around the park is doing to resident wildlife.

"This is the place where Theodore Roosevelt developed his conservation ethic, which went on to influence the history of the United States and certainly the National Park System," reflects James Nations, who leads the NPCA's Center for Park Research and edited a report on fracking. "The potential for fracking to symbolically roll over and surround the park is both ominous and disappointing."

Our wilderness areas and National Parks urgently need protecting

How will Washington ever be convinced to provide the funds necessary to make the backlog of repairs urgently needed in our wilderness areas/National Parks? And with Republicans still in control of the Senate, meaningful legislation on a host of issues is certain to be blocked. Advocacy groups can only do so much; our citizens need to become more involved, and mount public opposition to these destructive policies.

A comprehensive Green New Deal that draws together proposals for protecting our environment and improving the lives of Americans in other areas, has the potential to attract a widespread coalition of support. This is based on multiple public opinion surveys which show that the nation's citizens are concerned about these issues.

The Republican party's return under Trump to its all-out assault on our natural heritage must be stopped.

We need to pay heed to what Theodore Roosevelt once said: "I recognize the right and duty of this generation to develop and use the natural resources of our land, but I do not recognize the right to waste them, or to rob, by wasteful use, the generations that come after us. Conservation is a great moral issue, for it involves the patriotic duty of insuring the safety and continuance of the nation."

7.
Saving the Forest for the Trees

IMAGINE THE MAGNIFICENT NATURAL FORESTS THAT GREETED THE first visitors to the land we call North America: Forests of giant maple, chestnut and oak stretched from the east coast to the Mississippi Valley. Spruce and pine carpeted the Great Lakes region. Ancient redwood and Douglas fir towered over the Pacific Ocean. Nearly half of our continent was covered with woods.

There were thousands of plants, animals, fungi and microorganisms living in dynamic balance—a fantastic interplay of the sun's energy, the earth's minerals and pure water creating a magic web of life. There were trees as tall as skyscrapers, animals fierce as the wolverine and powerful as the grizzly bear, birds as beautiful as the scarlet tanager and majestic as the bald eagle.

Our ancient woodlands have survived world wars, the gold rush and the industrial revolution. Yet now the U.S. Forest Service, Congress, and the President are their most deadly threats.

While 95 percent of our original forests were logged during the past 243 years, a few remnants of never-logged, virgin woodlands remain in pockets throughout America. It is here that the native species of trees, plants and animals can still be seen...for now. These pristine natural forests store a living library of forest life descended from times past. But they are in jeopardy.

Our wooded areas began their evolution eons ago. Some of America's forests are still home to trees 1,000 or more years old.

As the United States was born, and the population spread westward, trees were typically felled with axes. This was a tedious process, to be sure, especially when constructing even the most modest shelter such as a log cabin. Railings for fences also had to be cut to prevent livestock from wandering off.

With the arrival of more and more immigrants from overseas,

the demand for lumber resulted in much of our forestland east of the Appalachians being seriously reduced by about 1850. So New York lumber barons began to seek forested areas farther to the west. Most desirable were stands of old-growth trees—especially white and red pine, considered at the time to be most suitable for home construction. A lot of these trees were said to be 200 feet tall.

A couple of men using a "cross-cut saw" could fell even a large tree with a lot less work than doing so solely with an axe. The invention of the steam saw by the middle of the 19th century made this process even faster.

Investors bought private and government land, established saw mills, and set up lumber camps that were often comprised of around 70 men, who came to be known as "lumberjacks." They worked from sunrise to sunset, six days a week, and did most of their work in the cooler weather months, or even winter.

As Gregory Fournier explains, "A ten-man crew could produce about 100 logs a day with a two-handled, cross-cut saw and double-edged axes.

Logs were too heavy for men to lift, or to be carried by a wagon with wheels, so they were loaded onto a sled by being pulled up a ramp by a team of horses fastened to a pulley on the other side.

The loads were large! An 1893 issue of *Scribner's* magazine de-

Lumberjacks hard at work sawing a tree by hand in the winter.

(Top) Horses pull logs up a ramp onto a sled. (Bottom) Ready to go!

scribes a load of logs "18 feet long, 15 feet wide, and 33 feet, 3 inches from the top to the roadbed, weighing over 100 tons...hauled by a single team" over the specially prepared ice covered trail.

The horses were generally shod with special "clawed" shoes, which gave them a firmer footing on ice. The drivers had to be especially careful going down a grade as the load could overrun the horses. If they had an uphill grade they would add a couple of helper horses.

The load of logs would be dragged to a river's edge to await the arrival of spring.

A fellow known as a "lumber sealer" would mark the ends of the lumber with the "seal" of the company that was doing the logging. This was to differentiate the load from other logging companies using the same mills—much like branding cattle.

The logs were put into the river in the spring and floated to a sawmill. The lumberjacks made a water corral out of some of the logs by chaining them together end to end. The bulk of the logs were put inside the corral.

Steam tugboats pulled the log corrals down inland rivers and even out across the Great Lakes. The tugboats waited in an area where the logs would appear just like a taxi waits for a rider.

But there were problems. The Michigan Department of Resources states that "high quantities of sediment from log drives and sawdust from sawmills were dumped into rivers... Whenever sawdust was dispensed into the river, toxic and oxygen-deprived conditions were created for fish. These detriments, combined with land clearing efforts, exacerbated soil erosion into rivers, significantly reducing the quality of fish habitat in rivers."

Gregory Fournier says that wasn't the end of the problems. "When the logging industry was finished raping the land, lumber camps were abandoned because owners didn't want to pay taxes on the land they owned, so they simply defaulted and the land went to the state. In all, over 19 million acres were clear cut with no reforestation strategy, leaving behind barren 'stump prairies' contributing to soil erosion, river and lake pollution, more atmospheric carbon dioxide, and degraded wildlife habitat."

From about 1890 to 1940, over-logging, plus torrential floods washed away the rich topsoil and gouged deep ravines, exposing rocks, clay or sand. The bare earth was saturated with plant-killing chemicals, the land bulldozed, the remaining branches and fallen tree trunks soaked with gasoline and set on fire; every living thing above and below the ground was dead. The entire scene was crisscrossed with dirt roads leading to...nowhere.

Logging companies recklessly cut all but the most remote and steep slopes across the nation, to provide raw materials for houses and factories in a growing industrialized country. Eventually, the demand for wood slackened as steel became the construction material of choice for the military and other industries. By then, though, the notion of an old growth forest existed mainly as a fading memory. The record of that devastation is available on microfilm in newspaper archives and in old books.

Despite the visionary leadership provided by President Teddy Roosevelt, an ardent conservationist—who set aside 150 million acres of wilderness for future generations, and created the U.S. Forest Service—our native woodlands continued to be plundered without reseeding.

Teddy's cousin, Franklin D. Roosevelt, reflected on this when he spoke at a dedication ceremony for the Great Smoky Mountains National Park on September 2, 1940. FDR commented, "In these centuries of American civilization, greatly blessed by the bounties of nature, we succeeded in attaining liberty in Government and liberty of the person. In the process, in the light of past history, we realize now that we committed excesses which we are today seeking to atone for.

"We used up or destroyed much of our natural heritage just because that heritage was so bountiful. We slashed our forests, we used our soils, we encouraged floods, we overconcentrated our wealth, we disregarded our unemployed—all of this so greatly that we were brought rather suddenly to face the fact that unless we gave thought to the lives of our children and grandchildren, they would no longer be able to live and to improve upon our American way of life."

The devastation of our forests continues

Clearcut logging is the complete removal of all trees in a given area, a practice that can be devastating to people and animals alike. The Natural Resources Defense Council has described clearcutting as "an ecological trauma that has no precedent in nature except for a major volcanic eruption."

Here are some scenes from our National Forests:

• Bald Eagles flying in terror as their 400-year-old nest trees crash to the ground.

• Salmon and their eggs smothered under an avalanche of mud sliding off of a clearcut mountainside.

• Bears fleeing as chainsaws and bulldozers clearcut their forest sanctuaries.

• It's not just the big timber companies at fault. With a wood-cutting permit in the Klamath National Forest, John Q. Public can go out in the woods with a chainsaw and cut down a 100-year-old oak tree, or cut all of them from a hillside for that matter, as long as he has paid his $20 for four cords of wood.

• Even the Giant Sequoia National Monument, home to the world's oldest and biggest trees, is being illegally logged by our own government, a practice that has continued under various Presidents.

These are not exceptions. For decades, the U.S. Forest Service has deliberately clearcut millions of acres of public woodlands, selling off the logs to international timber corporations at bargain basement prices. Clearcutting describes completely clearing an area of trees, regardless of their size and usability. Remaining scrub and brush are usually burned in large piles that can cast a smoky haze over the countryside for weeks. A clearcut area may be relatively small, or may span miles, and despoil the landscape—including the scars of logging roads cut to access it. The abrupt removal of trees can have a serious environmental impact on the surrounding area.

For instance, clearcutting may profoundly alter local rivers. If logging comes close to the banks of a river, as it often does, it eliminates the shady shield of trees, which can cause the temperature of the river to elevate. Even a few degrees can make a huge difference to native plants, fish, and amphibians, and can cause a significant population decrease. Numerous organizations monitor global rivers and have warned that extensive clearcutting could result in the extinction of some fish species, as they are driven out of their native habitats. Clearcutting also softens the banks of the river by enabling erosion, which can cause them to collapse into the water.

Here's what a clearcut forest looks like after the U.S. Forest Service and/or timber companies get through decimating it.

In a recent ten-year period, our Forest Service spent more than $5 billion in tax dollars carrying out the destruction of our National Forests. Timber barons rake in millions, while American taxpayers lose billions.

National Forest logging contributes only three percent to the nation's timber supply and the Forest Service actually loses over a billion dollars a year on logging. Logging National Forests puts people out of work by displacing logging on private lands and by competing against producers of recycled or non-wood fiber and building materials.

There is a world deforestation crisis, and our country is at the heart of it.

According to the Center for Public Integrity (CPI) based in Washington, DC, "For years the [U.S. Forest Service] has been one of the most mismanaged, poorly led, politically manipulated and corrupt agencies in the federal government." It said it is "an agency that is at war with itself and nature—one that mistreats and muzzles its own employees, routinely breaks the law, places its own interests over its mission to care for the land and in general sleeps with the [timber] industry." It also concluded that Congress' legislative agenda in regard to forests is "substantially set by timber industry interests."

This affects citizens at every level, from campers, hunters, hikers, fishermen, and bird watchers to cities and towns that rely on clean, mountain-fed drinking water.

The U.S. Congress has directed the Forest Service to do this clear-cutting on behalf of wealthy and powerful timber and paper corporations. These corporations have bought political influence with Congress to pass laws that liquidate our priceless public forests for private profit. Clear-cutting promotes overuse of paper and wood, clogging landfills and filling incinerators with waste.

After the Forest Service clearcuts our National Forests, in their place it plants tiny, identical tree seedlings-tree farms.

Tree farms contain only one species of tree planted by the thousands in row after row, all the same age. Having none of the diversity and fertility of natural forests, tree farms are prey to destructive fires, drought and floods, and invasions of pests and diseases. On millions of acres, tree farms have failed to grow, leaving huge regions of our National Forests permanently barren.

This subsidized destruction of our National Forests leads to a chain reaction of deforestation and environmental pollution around the world.

The U.S. government's dumping of cheap, subsidized public timber artificially lowers wood prices and devalues private timberland. Small, private woodlot owners can't match subsidized government prices, and are forced to overlog and clearcut to compete in the marketplace. The U.S. exports much of its timber overseas as raw logs and wood pulp for foreign corporations. Meanwhile, we import timber from tropical rainforests, causing massive deforestation around the world and loss of jobs here at home.

The billions of dollars currently spent subsidizing the logging of public lands could instead employ tens of thousands of people to restore forests rather than destroy them.

How It Used To Be

America was once covered with one billion acres of towering primeval forests. These forests were teeming with plants and animals, a treasure-trove of evolutionary diversity and biological richness. Giant, centuries-old trees had trunks more than 15 feet wide and soared to the height of 30 story skyscrapers.

In the past 500 years, aggressive logging and development have destroyed most of these original forests. The last remnants of America's virgin and natural woodlands, with their unique and irreplaceable life, reside mostly in our National Forests. These 155 forests cover a large portion of our country—an area about the size of California, Oregon and Washington combined, and stretch from Alaska to Florida. Most states have at least one National Forest.

These timberlands provide important biological services that most of us take for granted. Forest ecosystems recycle nutrients, wastes and produce soil. They play a vital role in mitigating changes in our global climate by absorbing and storing vast amounts of carbon from our atmosphere (53 million tons per year). National Forests also offer pest control and pollinating services by providing habitat for species that prey on forest and agriculture pests and for wild pollinators that are instrumental in the survival of certain crops. Forests also contain life-saving medicinal plants, including two important cancer-fighting agents that have been found in trees growing on undisturbed National Forest lands.

Deforestation is occurring on a massive scale in our National Forests and is clearly visible from space. Satellite photos show that the rate of clearcutting in places like the Olympic National Forest of Washington state equals or exceeds the destruction in the Brazilian rainforests. A World Resources Institute report concluded that the last of

Trees toppled over in this mudslide.

the original forests in this country will be lost without immediate action.

For decades, citizens have attempted to stop Forest Service destruction of our public woodlands by using the timber sale appeals process, lawsuits, and participation in national forest planning. Despite all these efforts, the Forest Service continues to allow private timber companies to clearcut old growth and roadless forests throughout our National Forest system, destroying critical habitat, ruining important recreational areas and violating the public trust.

The costs to the American people in environmental damage and wasted tax dollars are staggering. Increased species extinction, flooding and landslides are examples of the destruction resulting from clearcutting in fragile forest watersheds.

Natural forests act as giant sponges that regulate the flow of water into streams and rivers. During and after rain, the trees and shrubs hold vast amounts of water in their trunks and leaves, and their roots bind and stabilize the soil.

Clearcut areas don't absorb water. Instead, when heavy rains come, they allow for rapid runoff, causing flooding and erosion. The floodwater transports tons of silt, clogging waterways. In steep areas, the earth can no longer resist the tug of gravity and pulls away in a landslide.

Downstream in the valleys, homes and lives are ruined by a wall of water and mud. Government subsidies are needed to help communities and individuals repair the damage. In recent years, major floods and landslides in California, Oregon, Washington and Idaho have caused billions of dollars in damage to public and private property. Many people were injured and some were even killed. Some landslides were directly attributable to clearcut forest areas.

This tragedy of deforestation on our public lands is multiplied by the fact that taxpayer dollars help subsidize the clearcutting of our national woodlands. Billions of dollars are allocated to the Forest Service to pay the costs of building logging roads and administering timber sales. The timber industry buys the subsidized timber. The result is private profit causing public forest destruction.

The Orion North timber sale of old growth trees in the Tongass in southeastern Alaska exemplifies the stupidity of the U.S. Forest Service timber sales. Fortunately, it was halted by a federal judge in December 2009, after Earth Justice brought the issue to court. The project would have cost taxpayers close to $1.6 million while only generating $140,635 from the trees. Meanwhile, old growth trees are cut, a beautiful forest ecosystem is devastated, and new roads are pushed in a pristine area. One wonders: Is this just to create jobs and to keep mills open? Is the Forest Service a human welfare organization? Are they even concerned about the health of the forest ecosystems?

Natural forests are home to thousands of native plants and animals interconnected in a delicate web of life. Each organism is interdependent on the other. The spotted owl eats voles, a small rodent. Voles eat fungi and disperse the fungi spores in their waste which then grow in the ground on the roots of the giant trees. The fungi are essential to helping the trees take up vital nutrients through their roots from the soil. Each organism plays a role in the healthy functioning of the forest.

The forest is teeming with life, from common insects living in rotting logs on the forest floor to rare moss and lichens that only grow in the branches of trees, high in the forest canopy.

Because of massive forest destruction caused by clearcutting, the delicate web of life in our forests is unraveling. Scientists say that the earth is experiencing a wave of extinction. The leading cause of extinction is destruction of native habitat by such human activities as clearcutting and logging roads. Each year of continued clearcutting in the national forests leads to the loss of more species.

Millions of acres of these spectacular forests in the Pacific Northwest and the Sierra Nevada desperately need to be protected for future generations. These forests of thousand-year-old trees are home to the endangered spotted owl, the marbled murrelet, fishers, martens, and other animals emblematic of America's wildlife heritage.

The lifeblood of these pristine forests are crystal clear streams which provide habitat for endangered salmon and trout. These ancient forests are faced with the constant threat of logging, placing many species in danger of extinction.

Even our biggest trees are threatened

The sequoias are among the largest, longest living trees on earth. These ancient sentinels grow in a limited area of the Sierra Nevada Mountains, confined to about 150 secluded groves. Although some of these groves are protected within National Park boundaries, many of the groves lie in the National Forests—unprotected.

Redwoods are often confused with sequoias but they are very different trees. The wood of each may be red, and the cones may both be small, and both have very tall examples. But redwoods are coastal— primarily northern California. Sequoias are inland.

At 210,000 acres, Nanning Creek Grove, one mile east of the logging town of Scotia, CA is the largest unprotected stand of primeval redwood forest in the world. It contains old-growth trees up to 15 feet in diameter standing over 300 feet tall, likely to be thousands of years old.

In 1986, Houston financier Charles Hurwitz, helped by former

Looking up at these towering giants is an awe-inspiring experience.

junk bond titan Michael Milken, purchased Pacific Lumber in a leveraged buyout. In turn, Hurwitz greatly increased Pacific Lumber's harvest of redwoods in order to meet annual interest-only payments on $850 million in junk bond debt, a figure that later grew to $950 million.

The timber harvest plan for the area was nicknamed "Bonanza" by Pacific Lumber, which expected to reap a financial windfall from felling the massive trees.

Darryl Cherney, a former Earth First activist who played a prominent role in the drama, told the local *Press-Democrat,* "I was flabbergasted that ancient redwoods were being logged. I didn't know it was legal."

In 1990, a car bomb blast nearly killed Cherney and Judi Bari. The pair were driving in Oakland, recruiting volunteers for protests they dubbed Redwood Summer, when a bomb exploded under the front seat of their car, shattering Bari's pelvis. This attack brought international attention to the timber war.

Other high-profile events and deaths followed over the years.

By 2007, Pacific Lumber had declared bankruptcy and the court approved a deal the next summer which brought Humboldt Redwood Co. (HRC) into the picture. Except this time, a man named Sandy Dean, who headed the new company, vowed to protect the endangered trees.

John Andersen, director of forest policy for HRC, explained in a 2019 interview for this book that "When we purchased the property, there was a treesitter in one of the largest old growth trees in this grove. One of [our] policies is to retain all old growth found on the forestland, whether it is an individual tree or a larger stand of trees. When the treesitter and other environmentalists heard of the new policy they came down from the tree and were happy to hear of this new policy. The Nanning Creek Grove still stands as it did in 2008 and our policy of retention of old growth trees has remained the same."

Andersen says that "Humboldt Redwood Company was formed in 2008 with the purchase of 203,000 acres of redwood and Douglas fir forests with the intent to manage the lands with a high level of environmental stewardship while operating as a successful business... We selectively harvest (thin) our forest to supply logs to our sawmills in Scotia and Ukiah, CA. To ensure we are operating with a high level of enviormental stewardship we [sought] third party certification from the Forest Stewardship Council, who certifies forestland owners globally when they meet strict criteria, including operating sustainably. We were successful in achieving this certification and have remained cer-

For decades, Americans have thrilled at being able to drive through tunnels in these behemoth redwood trees along the "Avenue of the Giants" in northern California.

tified since this achievement.

"The history of our land has been one of over harvesting, so we harvest less than we grow each year to rebuild the forest to its former glory. But that takes time," he admits. "It takes decades, around six decades, to grow a redwood from seedling to merchantable size. The diameter of our redwood only grow 1/2-inch to 1-1/2-inch per year so our investment into this forest is certainly a long term one. We do not grow Sierra redwood on our property as one of the conditions of our certification is that we plant only tree species native to the property. As we are located in the belt of coastal redwood, Sierra redwood would be a non-native species in our portion of the state."

The fight to protect the huge trees of California's Sierra Nevada range began in the late 1800s, when Sierra Club founder John Muir won the establishment of Sequoia National Park.

In April 2000, at the Trail of 100 Giants in the Sequoia National Forest, President Bill Clinton announced the creation of the Giant Sequoia National Monument. The 328,000-acre area is intended to protect 38 groves of these giant sequoia trees, found only on the west slope of the Sierra Nevada.

True to form, George W. Bush attempted to log them, but in 2006, Judge Charles Breyer blocked him. California Attorney General

Bill Lockyer said, "The ruling...is a resounding defeat for the Bush administration, which aggressively sought to unravel the protections guarding John Muir's big trees. We simply cannot afford to sacrifice for short term gain majestic sequoias that have stood for centuries." (The trees can live to be 3,000 years old!)

Yet these groves of towering trees are still threatened. The Forest Service has called for extensive logging of this natural cathedral, under the guise of fire protection. The Forest Service's own scientists have found that logging large, fire-resistant trees like those in the Monument does little to prevent catastrophic wildfire.

According to the website *Beyond Ordinary*, "You wouldn't know that sequoias' orange, spongey bark is relatively fire resistant. It feels vulnerable to the touch and a knock to the trunk sounds hollow. But these giants move forward—they grow up and they grow out—when fire licks the ground around them. Their scarred trunks are testaments to the the thousands of years they've witnessed on earth, and their ability to use the fire that weakens their surroundings is what helps them grow to be the largest trees in the world.

"The biggest trees in Kings Canyon and Sequoia display deep cut burn scars at their bases that take years to heal. It's a beautiful parallel to our own human existence—we experience trauma, hurts, and loneliness, and time, above all else, helps us to heal. Too many fires close together and like the sequoias, we break, but with a little luck, our bark has had enough time to cover the last scar before a new one cuts into us. And no matter what, we move on—we grow up and we grow out—because it's what we have to do."

Beyond Ordinary points out that "Sequoias only started growing on earth about 4,000 years ago, a relatively new development considering the history of the world; about 400 years before ancient Greeks started to flourish and about 700 years before Moses led the Israelites out of Egypt, for context. Their trunks grow out after they reach a [height of] 275 feet or so, making it a grecian pillar, the same width at the bottom as at the top, towering over the landscape as a living statue. Their scale stops you in your tracks and draws your gaze upward to their mounded tops, a sign of their maturity, and one of the many characteristics that makes them a unique tree. We've all heard about the benefits of being in nature and getting outside, and a trip to the sequoias offers all of that, in addition to a reminder of how to live.

"Evolution allowed for these awe inspiring giants to grow in the very specific conditions of the Sierra Mountains in California, during a time period when it was just warm enough and just cold enough and with just enough rainfall, but not too much, at just the right elevation. They are simply growing into their design, and being in their presence is a reminder for us to do the same."

In a typically misguided policy, the U.S. Forest Service has riskily allowed logging all around these fragile groves. The sequoia grows naturally in stands of trees with many other species, such as Ponderosa pine and Douglas fir. Logging within the sequoia groves threatens the existence of the sequoia. It exposes the trees to the full force of the wind and kills the intertwining root systems that help keep the sequoia standing.

Two areas are especially vulnerable

The Rocky Mountain region has the largest unroaded areas in the lower 48 states. Many roadless forest areas in this region remain mostly as they were before Columbus set foot on this continent, wild and untamed. This is the only region in the lower 48 states where the grizzly still roam free. Massive new logging and roadbuilding projects in these National Forests are already degrading this region. With light rainfall, mountainous terrain, and fragile soils, these forests are particularly vulnerable to clearcut logging. When timber companies clearcut these forests, they may not grow back for centuries.

Like the giant sequoia groves, the Sipsey wilderness of Ala-

Turkey Foot Falls in the Sipsey Wilderness, Bankhead National Forest.

bama, and the Cochetopa Hills of Colorado are threatened by clearcutting and roadbuilding.

The Sipsey wilderness is located in the Bankhead National Forest in Alabama. The Sipsey is a wonderland of diverse hardwood forests dissected by mysterious canyons harboring rare plants and secret waterfalls. Noted for its outstanding biological diversity, half of all fern species in Alabama are found in the Sipsey along with 147 species of birds and 53 kinds of amphibians and lizards. Migrating songbirds find rare habitat in the interior forests.

Flowing through the wilderness is the Sipsey River. It is designated a Wild and Scenic River, and is home to an endangered species of freshwater shellfish.

The U.S. Forest Service has allowed destructive clearcutting throughout most of the Bankhead National Forest, degrading the entire ecosystem and threatening the survival of the Sipsey Wilderness area itself. The Forest Service has been destroying the diverse species of plants and animals in this woodlands as well, and replacing these beautiful natural forests with sterile tree farms.

The forests of the Cochetopa Hills are spread throughout three National Forests in the southern Rocky Mountains of Colorado—the Gunnison, the Grand Mesa and the Uncompahgre. The Cochetopa Hills are known for unusual mixed stands of conifers. Ponderosa pine, Bristlecone pine, aspen and Engleman spruce are found growing together. This is just one aspect of the Cochetopa Hills' high biological diversity.

Cochetopa means "pass of the buffalo" in the Ute language. This descriptive name reveals the unique quality of the area as an ecological interface zone and important wildlife corridor. The pass at Cochetopa is low elevation, creating vital winter habitat and migration corridors for many animals such as black bear and elk. Distinctive interior wetlands also attract wildlife to this semi-arid part of the Rockies. These rare wild forests are threatened with destructive logging and roadbuilding.

In 1995, the timber industry pressured Congress to pass a bill that suspended all environmental laws on the National Forests for two-and-a-half years, and increased logging of rare, previously protected Ancient Forests. The Ancient Forests, the roadless areas forests, and other biologically critical forests were clearcut and degraded, and the public could not stop this destruction.

There continues today to be a concern about forest management practices on both federally held and private and industrial lands.

To many recreational users of National Forests and to conservationists, the U.S. Forest Service has been more interested in timber sales than in managing the forestland for wildlife habitat and human recreation.

The paper industry harvests forests and sells to foreign countries

The American paper industry continues to whittle the forests of the Southeast away, shipping the region's trees down the railroad tracks and rivers for short-term profits.

Many of the "chip millers" are selling the South's renewable forest resources down the Tennessee-Tombigbee waterway to Korean and Japanese companies to make computer printer and high-quality magazine paper to sell back in the U.S. market. This registers as a net loss on the International Trade balance sheets, and as a loss to local economies, especially for the local saw-timber folks who have trouble competing for trees to cut.

Several forest types of national and global significance are found in the 13 states that comprise the American Southeast, in some of the richest temperate forests on earth. Along with the streams that flow through them, they are renowned for the diversity of native plants, fish, mussels, amphibians and migratory birds, including some threatened and geographically-restricted species at risk of extinction.

How times have changed!

Instead of two-man teams cutting trees from sun up to sun down with giant cross saws as in the old days, this time around, the ef-

Modern timber cutting is cruelly efficient.

ficiency of the machinery for felling trees is unparalleled, like something out of an Edward Abbey novel. Mammoth bulldozers drag whole trees on cables up the steepest of mountain slopes. It's called cable logging, and you have to see it to believe it. On one hand, logging methods today are technological, industrial marvels—highly productive and cheap. For instance, one company's investment in East Tennessee land and a chip mill totals $13 million, and it takes only four to six men to run it. Much of the logging itself is hired out to local contractors who must submit competitive bids. The company employs a grand total of 12 people in Tennessee, about one per $1.1 million investment.

While the wood products market is a bit volatile and prices vary, estimates place the average value of a saw timber tree at $400. But when taken for chips, the smaller, lower quality trees go for about $4, from land where property taxes average only $1.50 to $2 an acre. It's quick and easy to turn a forest into chips. The size of the log doesn't matter, and the transportation method is cheap. All of it amounts to what environmentalists are calling a "quiet rape" of the Southeastern environment.

Despite the existence of alternative fibers for pulp, such as hemp and kenaf, about half of the trees cut each year are turned into paper products. Fifty percent of the landfill waste in America is wood and paper fiber.

There are nearly 150 chip mills operating in the region, 100 built in just ten years, with more proposed or currently under construction. Each high-capacity mill can go through 10,000-acres per year at maximum production. Best estimates suggest that clear-cutting to supply them consumes a total of 1.2 million acres a year. Concerns involve the harvest of new species and smaller trees, the loss of mature trees, impacts to threatened and endangered species and migratory birds, the loss of wildlife habitat and reduced water quality due to soil erosion from clear-cutting.

Midwest forests aren't spared, either

Most people probably don't think of the North Woods of Wisconsin as holding significant plant and animal species—but they do—including the goblin fern, American ginseng, the goshawk and the pine marten. These northern Wisconsin timberlands also hold some of the only remaining swaths of old-growth trees in the Midwest and provide a connection to nature for thousands of people in the region. For years the U.S. Forest Service has been proposing increased logging, timber sales and road building. But these North Woods are already the fifth most-heavily logged National Forest in the country and the most-heavily logged in the eastern half of the United States.

Don Waller, University of Wisconsin botany professor, says "This area is gaining in biological value right now. We have gray wolves coming back down from the Upper Peninsula [of Michigan]. We're beginning to hear of sightings of cougar. We're seeing a recovery of bird species. Will this recovery be able to continue?"

Bush tried to help timber barons

George W. Bush's administration was famous for launching programs that had a deceptive name—like "Healthy Forests Initiative"—when in reality, their goal was to do the exact opposite of what was claimed. Such deceit of the public was routine for Bush, et al.

What was actually his "leave no road-builder behind" policy—a pander to timber barons—is especially evident in how he sought to undermine the Forest Service's "roadless rule," a Clinton-era regulation that protected federally-owned forests not already crisscrossed by more miles of roads than are included in the Interstate highway system.

When enacted by Clinton, the roadless rule got more public support—over a million supportive messages came in to the Forest Service—than any regulation in history.

Mr. Bush got some help from a key Democrat, Senator Tom Daschle of South Dakota, who deviously slipped language into a bill in the summer of 2002 that allowed logging on Native Americans' land (and other areas) without having to abide by environmental restraints or lawsuits. These very holy lands were once visionary refuge for Lakota Sioux elders including Crazy Horse and Black Elk.

As environmental writer Jeffrey St. Clair wrote at the time, "The logging plan was consecrated in the name of fire prevention. The goal of the bill, Daschle said, 'is to reduce the risk of forest fire by getting [logging] crews on the ground as quickly as possible to start thinning.' It's long been the self-serving contention of the timber lobby that the only way to prevent forest fires is to log them first.

"The legislation however, accomplishes no such thing, and instead sanctions the pillage of over 2.5 million acres of federal forest land by 2012."

So Bush got rid of the roadless rule, and millions of acres of roadless forest were opened to road-building, clearing the way for lumber, energy, and mining corporations to get in and take what they wanted. The impact was even greater in the wilds of the Northwest where new roads are sure to aggravate the silting up of streams and rivers in which depleted stocks of salmon are struggling to hang on.

But then Bush hit a roadblock. On September 20, 2006, U.S. Magistrate Judge Elizabeth Laporte ruled against the Bush administration's plan to reverse the Clinton-era regulations, saying that the Bush plan "established a new regime in which management of roadless areas within the National Forests would, for the first time, vary not just forest by forest but state by state. This new approach raises a substantial question about the rule's potential effect on the environment."

On November 29, 2006, Judge Laporte even issued an order to ban road construction on 327 oil and gas leases issued by the Bush administration since January 2001, most of them in Colorado, Utah, and North Dakota—areas that were already protected before the Bush reversal of the 2001 law.

On May 28, 2009, during the Presidency of Bush's successor, Democrat Barack Obama, Secretary of Agriculture Tom Vilsack issued a directive giving the Ag department final authority on most road development and timber activity in National Forests, for a period of one year.

In 2011, a federal appeals court in Denver, Colorado upheld the government's authority to prohibit Western states from building roads on public land. The unanimous ruling, issued by a three-judge panel, said a lower court had erred in finding for the State of Wyoming, the plaintiff in the case, and ordered that the rule be put into force nationally. It was poetic justice that the decision was written by Judge Jerome Holmes, who was nominated to the court by President George W. Bush.

As of this writing the roadless rule is still the law of the land after surviving its final legal challenge on March 25, 2013, when the U.S. District Court for the District of Columbia rejected the state of Alaska's challenge that, while aimed at the Tongass National Forest, would have nullified the national rule. The Alaska case was the final litigation challenging the rule nationwide. The Court held that no further challenges are allowed, because the statute of limitations has run out.

But now that Donald Trump is President, the United States Department of Agriculture (USDA) announced in August 2018 that it will develop a state-specific roadless rule focused on the Tongass National Forest which it feels "would require a different management designation to further Alaska's *economic development or other needs*," while still purporting to conserve roadless areas "for generations to come."

Citizen activism can counter exploitation of our forests

In 2004, Greenpeace began actively campaigning against Kimberly-Clark, which is the largest tissue maker in the world. Headquartered in Dallas, Texas, the tissue giant owns the Kleenex, Scott, Viva, Cottonelle, Kotex and Huggies brands. Kimberly-Clark produces more than four million tons of tissue products annually and generates billions of dollars of annual sales from 150 countries around the globe.

Unfortunately, Kimberly-Clark was cutting down ancient forests using clearcut methods in order to produce its tissue. Most of the pulp Kimberly-Clark used for its disposable tissue products came from unsustainable sources. This includes wood fiber from ancient forests like the Canadian Boreal. Most of the consumer products Kimberly-Clark had been selling in stores, including Kleenex products, contained no recycled fiber at all.

According to a study titled "The Carbon the World Forgot," boreal forests contain 703 gigatons of carbon, while the world's tropical forests contain 375. The boreal needs to be maintained, but for years Kimberly-Clark's clearcut logging practices did just the opposite.

Clearcutting is cheaper for logging companies, which means less overhead and more profit for corporations.

Of course, this is no excuse for Kimberly-Clark to support unsustainable practices from loggers that auditors discovered leaving more than 328 million cubic feet of trees cut and left to rot on logging roads in the Kenogami Forest of Northern Canada.

On November 3, 2005, a grassroots campaign culminated in the "Boreal Day of Action," with 350 events in 200 cities spanning the U.S. and Canada. There were demonstrations outside Kimberly-Clark facilities across the globe, including in New York City, Toronto, London, and Hamburg.

An advertisement was even published in *The New York Times*, attacking Kimberly-Clark in its most vulnerable place—their flagship brand, Kleenex.

As Kimberly-Clark refused to budge, the movement to stop deforestation gained in momentum and allies. Greenpeace started the Forest Friendly 500 initiative, asking businesses to pledge to stop wiping away Canada's boreal forests by replacing their Kleenex brand facial tissues with a different brand that uses more recycled tissue.

Activists continued direct action; they blockaded Kimberly-Clark mills and staged events in front of various Kimberly-Clark buildings. In 2009, Greenpeace released the "Recycled Tissue and Toilet Paper Guide," which was later complemented with an app.

After several years of effort from thousands of supporters across the globe, Kimberly-Clark outlined a sourcing strategy that makes it an industry leader in sustainable logging practices. It set a 2011 goal to have 40 percent of the fiber in North American tissue products either recycled or certified by the Forest Stewardship Council—a 71 percent increase from 2007 levels.

More importantly, Kimberly-Clark issued a demand to logging companies for sustainably logged outpaced production. This meant logging companies themselves were forced to shift toward more environmentally friendly forms of acquiring timber, leading to less deforestation in the entire industry. It set precedent for a higher standard of sustainable logging, while sending big business a message that consumers are willing to choose with their wallets when it comes to protecting the environment.

What happened in this case demonstrates how citizen activism can make a big difference!

Kimberly-Clark has even benefitted financially. The Scott Naturals line, with 20 percent recycled fiber and launched in part as a response to Greenpeace's efforts, has grown from $4 million in sales when the deal was reached in 2009 to $100 million now.

Other groups and businesses are taking the initiative to protect our woodlands as well. In 2018, American Forests and Alcoa Foundation announced nine new grant recipients in year two of a three-year partnership to enhance biodiversity and combat climate change in key areas across the globe.

Newly funded projects will restore at-risk tree species and help cities address climate change by engaging 4,500 volunteers in tree plantings and other forest restoration activities in 13 locations.

In Indiana, Friends of Patoka River National Wildlife Refuge and the U.S. Fish and Wildlife Service will restore 55 acres of bottomland hardwood forest on former agricultural land in Patoka River National Wildlife Refuge. A variety of oak and hickory species will be planted that will benefit federal and state wildlife species of concern including evening, Indiana, and long-eared bats, barn and short-eared owls, and the copperbelly water snake.

In Washington and Oregon, the U.S. Forest Service will collect whitebark pine seed from disease-resistant trees and use them to grow tree seedlings needed for future forest restoration. Whitebark pine is endangered in Canada and a candidate for Endangered Species Act protection in the United States, due to threats including a non-native fungal disease (whitepine blister rust), wildfire and climate change.

The timeless beauty of the forested Rockies continues to inspire.

Availability of disease-resistant whitebark pine seedlings is a key limiting factor to range-wide recovery.

These and other new projects add to ongoing restoration work previously funded through the American Forests and Alcoa Foundation Partnership in Iceland, Quebec, Canada and Pennsylvania, USA. Early efforts in 2018 planted 75,000 trees over 175 acres with the help of 2,500 volunteers.

Lewis and Clark would still recognize these forest lands

Mike Garrity—a fifth-generation Montanan and director of Alliance for the Wild Rockies—and singer-songwriter Carole King, who has lived in Idaho for more than 40 years, co-authored a piece in *The New York Times* in March 2019 about the threat facing the forests of their area.

They wrote, in part, "On June 12, 1805, Meriwether Lewis climbed a bluff in what is now Chouteau County, Montanna. From this vantage, he was able to see wolves, antelope, mule deer and 'immense herds of buffalo.' He added, 'From this height we had a most beautiful and picturesque view of the Rocky Mountains, which were perfectly covered with snow.' They were, Lewis wrote in his journal, 'an august spectacle.'

"Nearly 214 years later, parts of the Northern Rockies remain as they were when Lewis first saw its peaks. We must protect them so future generations can experience the grandeur that he beheld."

Mr. Garrity and Ms. King comment that "The Northern Rockies are surely near the top of the list of the world's most spectacular landscapes. Its ranges contain one of the last great expanses of biodiversity left in the continental United States, including most of the species that were there when Lewis and Clark first passed through in 1805 on their journey of discovery.

"These attributes alone would be reason enough to protect this region. Instead, the Trump administration has been pushing oil, gas, mining, and logging projects, and removing legal protections from threatened species. To be fair, the Obama administration also pursued some of those actions. But the current administration's zealotry threatens the region's wild landscape and rich biodiversity. It's up to all of us who care about the environment, science and preserving wild places for our children to resist such efforts."

Singer/songwriter Carole King is an environmental activist.

They write, "Legislation recently introduced in the House would protect a vast swath of this region. But until that law is enacted, we'll have to rely on the judiciary. Along with other organizations and Indian tribes in the Northern Rockies, our group, Alliance for the Wild Rockies, has been fighting threats to the region in court. Fortunately, this past year has brought some encouraging news. But the court system alone will not provide the protection this area needs and deserves.

"In August, a three-judge panel of the United States Court of Appeals for the Ninth Circuit voted unanimously to halt a planned 125-square-mile logging and burning project in the Payette National Forest in western Idaho. The court concluded that parts of the project ran counter to the forest's management plan.

"Under that project, so many trees would have been cut that the forest would have no longer provided elk or deer with the cover they need. Forest streams would have been filled with sediment from bulldozers building miles of new logging roads—further damaging the native fisheries for which the Northern Rockies are internationally famous. According to the United States Forest Service's own projections, taxpayers would have spent more than $12 million to subsidize the logging."

Mike Garrity and Carole King observed that "On the same day of the Ninth Circuit ruling against the Payette project, a Federal District Court judge in Montana, Dana Christensen, granted a request by Alliance for the Wild Rockies and two other groups for an injunction temporarily stopping a logging and road building project along the northwestern border of Yellowstone National Park. Nearly 16 miles of logging roads would have been bulldozed through grizzly bear habitat. The judge concluded that the Forest Service had failed to consider how the project would affect the Canada lynx, which has been listed since 2000 as a threatened species under the Endangered Species Act. Now the court is weighing whether to permanently block the project.

"Then, in September, Judge Christensen ordered the Yellowstone-area grizzly bears restored to full protection as a threatened species. When the federal government removed the bear's protected status in June 2017, ignoring the concerns of scientists, environmentalists and tribal leaders, the State of Wyoming started gearing up for its first grizzly hunt in more than 40 years. But Judge Christensen ruled that the United States Fish and Wildlife Service had erred in re-

When Europeans first arrived there were over 50,000 grizzly bears in the West. While 37 separate grizzly populations were identified in the contiguous U.S. in 1922, only six populations remained by 1975 when they were listed as threatened under the Endangered Species Act. Now there are only two areas in Montana with substantial grizzly populations—the Greater Yellowstone region with about 700 bears and the Northern Continental Divide region with about 900 bears. The isolated Cabinet-Yaak grizzlies number fewer than 50 bears.

moving the bear's threatened status, siding with us and opponents of the hunt. The Interior Department is appealing this decision.

"This was a particularly important ruling," the pair explain, "because the loss of protected status could have opened the bears' habitat to mining, logging, and development. But the fight to protect the landscape and its species is far from over. With the nation's environmental laws under all-out attack by the Trump administration, victories in the judicial branch can get us only so far.

Carol King and Mike Garrity assert "That's why Congress must act to protect the Northern Rockies. Legislation introduced recently...would designate 23 million acres of roadless public lands in Montana, Idaho, western Wyoming, and eastern Oregon and Washington as wilderness.

"The Northern Rockies Ecosystem Protection Act would also establish a system of vital biological corridors connecting smaller ecosystems within the Northern Rockies to protect native plants and animals. Restoring over one million acres of damaged habitat and watersheds would create jobs, and taxpayers would save millions of dollars that would otherwise be spent on road-building and logging projects in which private corporations cut down our public forests."

Various green groups have been fighting for more than a quarter of a century to protect the ecosystems of the Northern Rockies that are the rightful heritage of all Americans and generations to come. "With the support of an engaged public and insistence by an independent judiciary that this administration adhere to longstanding environmental laws, we will succeed" Gerrity and King concluded.

Opponents complain enactment of any of this legislation would mean a loss of extraction jobs in the northern Rockies: logging, mining, oil/gas production. This is a short-sighted and selfish argument.

The U.S. no longer leads the world in conservation

In 1872, when Congress designated Yellowstone as the world's first National Park, America was the world leader in conservation. That reputation is fast eroding.

In North America there are four wildlife corridors that have been proposed by the Wildlands Network, each providing a highway, called a "wildway," for migrating creatures to mitigate the effects of climate change: the Pacific Wildway running from Baja to Alaska, Boreal Wildway running west-east from Alaska, through Canada, to the northeastern shores of North America, the Eastern Wildway running from Everglades in Florida to the Arctic, and the Western Wildway also called the "spine of the continent" runs from southern Mexico along the Rocky Mountains up into the Arctic. The Yellowstone to Yukon Conservation Initiative (also known as Y2Y) is a bi-national NGO that promotes the conservation of habitats and wildlife movement ability from the Greater Yellowstone Ecosystem to the Arctic Circle.

A coalition of groups are trying to save the forest for the trees

American Forests, as the nation's first forest conservation organization, is taking charge and leads a 37-member Forest-Climate Working Group (FCWG), the nation's only sector-wide coalition working on forest-based carbon mitigation. Its president, Jad Daley, founded the Forest-Climate Working Group in 2007 and has served as co-chair since that time. FCWG represents a diversity of interests across the forest industry: from landowners to forest products companies, to the government, academia, and conservation groups. FCWG helps states identify how to use forests as a climate solution, this is achieved through their new toolkit.

Reforestation efforts, like the AF American ReLeaf program, are the single largest pathway to greater carbon capture. These efforts maximize carbon sequestration and long-term forest resilience.

Their work with the Northern Institute of Applied Climate Science is an unprecedented effort to identify how forest restoration and sustainable management can trap more carbon in forest soils—the single largest potential pool of carbon in forests. This will fill a longstanding gap in understanding of potential carbon capture in forests.

Tree planting in cities, through the Community ReLeaf programs, also helps absorb carbon, and creates even greater carbon savings by cooling the surrounding environment, which reduces energy costs. American Forests' urban tree planting is carefully targeted to urban heat islands, where this cooling benefit will be at its highest. Trees properly placed can cool the air the same as ten room-sized air conditioners running 20 hours a day, reducing air conditioning needs by 30 percent. In neighborhoods with more tree canopy, air quality improves by as much as 15 percent, reducing the incidence and severity of asthma, especially in children.

In August 2018, American Forests became the first group to commit to the U.S. Climate Alliance's Natural and Working Lands Challenge, a pledge to secure natural and working lands as a resilient net sink of carbon.

The group opines, "Through all of this, we have learned one thing for sure: the world needs more trees and healthy forests if we are to address climate change and create a habitable world for future generations."

Climate change is inextricably linked to forests

According to American Forests, "We cannot solve climate change without forests to help capture and store carbon dioxide (CO2) emissions made by humans. Forests absorb up to a third of carbon dioxide, the main greenhouse gas responsible for the planet's warming. So planting more trees affords us more of a chance to deepen the 'carbon sink,' which is the ability to absorb carbon dioxide from the atmosphere.

"With planting, we must be wise custodians of the mature forests that are already serving to reduce the amount of CO2 in the air, as well as storing vast reserves of carbon. These forests are struggling to adjust to rapid changes in our climate.

"Tree rings, the layers are grown each year, provide a record of how trees are affected by changes in temperature and rainfall patterns, and through them and other methods, we can see that climate change is weakening—and even killing—our forests at an alarming speed. From high-intensity wildfires and declining forest health from pests and pollution, the effects of climate change are worsening the threats to forests and reducing forest productivity.

"This is a pivotal moment for our forests and the health of our planet."

Trump gets it wrong when it comes to forest fires

While California was battling at least 17 major wildfires in the latter part of 2018, many of which were among the most destructive in the state's history, President Trump blamed "bad environmental policy" for the fires' intensity and "poor forest management." Then he vindictively threatened to withhold federal aid to the state.

Char Miller, a professor of environmental analysis at Pomona College and an expert on wildfire policy, told the *Pacific Standard* that the President's claims had no foundation. "This is theater," he said. "But in effect, it sends a signal that the President, who took an oath to protect the federal domain, is unwilling to do so."

A fire scientist, Matthew Hurteau—an associate professor in the department of biology at the University of New Mexico—pointed out that a large portion that was burning was not even forest. In the southern part of California, in the Malibu area, it was grass and shrubs that were ablaze.

Hurteau explains that "Fire is an integral part of California ecosystems because it consumes dead vegetation, creates space for new plant growth, and helps limit the density of vegetation. It affects almost every vegetated part of the state, from the conifer forests of the Sierra Nevada mountains to the oak woodlands lower down and, in the valleys, the grasslands and chaparral.

"The severity of these fires is moderated by rain and snowfall. California's Mediterranean climate means that the state receives heavy precipitation for only a few short months in the winter, and this is all that the vegetation has to tide it over until the winter storms begin the next year. As the temperature increases in spring and summer and plants use up the water stored in the soil, the amount of water held in plants decreases, making them more flammable. Similar to firewood, the drier it is, the easier it burns."

The scientist elaborates that "Climate

(This page) 2018 California fires were severe.

change is causing warmer temperatures, which dry out vegetation more. It is also causing winter precipitation to fall over a shorter period and the length of the fire season is increasing. Vegetation in California is increasingly primed for fire."

He says that "The native chaparral shrublands are dominated by shrub species that evolved with fires occurring every several decades to every century or so. Human activity has shortened the time between fires, however, which kills native chaparral species before they can produce seeds. These frequent fires also allow non-native grasses to invade. Unlike shrubs, such grasses are capable of burning nearly every year and still recovering."

Hurteau concludes "Donald Trump tweeted that more 'forest clearing' was needed to stop destructive forest fires. In fact, clearcutting is not going to solve the wildfire problem in forests either. Cutting trees won't stop wildfires from occurring where there are no trees. We cannot cut our way out of this problem."

Miller agrees. He notes that environmentalists have actually linked clearcutting to some of the country's biggest fires. Emily Moon of the *Pacific Standard* adds that "Poorly managed clearcuts leave behind debris, which can fuel a blaze. Moreover, many of California's ongoing fires...are not burning in forests—they're in grasses or chaparral, which the state does not clearcut anyway."

Climate change has turned the state's fire season into a yearlong scourge. Extreme temperatures and prolonged drought, like the one southern California has been experiencing, have been linked to a feedback loop that intensifies fires' effects.

The fact that Trump refuses to acknowledge the role of climate change, Miller says, is "disheartening."

Char Miller says that a second factor contributing to fires is that over the last few decades, three to four million Californians have moved into rural areas—high-risk zones for fire, where a source as small as a spark from a car can set off a destructive fire.

"Where people move, fire follows," Miller says. "We have to figure out a way to better live in the ecosystems that we now call home."

Poor morale in the U.S. Forest Service itself

The visionary Theodore Roosevelt had created the U.S. Forest Service specifically to acquire and protect vast stretches of wildnerness for future generations. But by 1920, Big Timber had co-opted the federal agency, leading to industrial clearcuts that "scalped" the land. The Forest Service became the fire service, protecting trees so industry could cut them down later.

Susan Marsh, who spent three decades with the U.S. Forest Service and is today an award-winning writer living in Jackson Hole, Wyoming, writes that "For many years the U.S. Forest Service was considered one of the best federal agencies to work for. To those who don't know, the service originated as a conservation agency, devoted to watershed protection and sustainable timber harvest.

"Until the 1930s, management of our national forests was mostly custodial. Boundaries were established and marked, field stations constructed, and later, telegraph and telephone lines installed.

"Early rangers caught game poachers, timber thieves, rogue miners, farmers illegally using the forest for their businesses, among other activities. They lived and worked out of remote cabins, which is part of why the Forest Service became a decentralized agency with much discretion left to local rangers."

However, she says in 2019, "the morale inside many federal land management agencies is in a free fall."

Originally, the Forest Service was well-thought of by the public, largely for its success in reducing wildfires. By the 1960s, the acreage burned by wildfire had declined by 90 percent compared to the 1930s.

Over time, new and different kinds of uses fell under Forest Service management and the agency shifted focus in modern times. Demand for timber products after WWII was a primary influence, and the workforce swelled to respond.

"New employees coming out of forestry schools shared common values and a sense of purpose: they were supporting the post-war prosperity and progress of the nation," Marsh reflects.

"Timber production began a three-decade rise in 1940, topping 12 billion board-feet harvested between 1965 and 1975. During the same decades, Americans gained enough leisure time and prosperity to travel to the forests for recreation beyond uses of the backcountry by hunters, outfitters and dude ranchers."

Ms. Marsh feels that "Timber production and public use of the forests clashed in the 1960s and '70s as people seeking beauty found large clearcuts instead. Soil eroded into rivers and, as pioneering conservationist Bob Marshall, once a Forest Service employee himself, famously lamented, the nation's wilderness was disappearing as quickly as a snowbank on a summer afternoon. The Forest Service became seen by many as an advocate for the timber industry and not much else.

"An explosion of reforms, new mandates and laws followed.

What citizens wanted from their National Forests changed, and the old-line foresters now had to share their work space with wildlife biologists, soil scientists, and archeologists, many of whom were of a younger post-Earth Day generation.

"Gone were the days when everyone within the agency thought alike, and gone were the articles of praise by the press. Issues became more complicated as our understanding of nature grew beyond viewing the land only for its commodity exploitive value."

She explained that sometimes there would be an infusion of federal funding for a few years, whereby maintenance in the National Forests was brought up to date. "Then the money dried up and it hasn't come back," she laments. "Twenty-five years later, with no funds for maintenance, the roads have gone back to their earlier condition and you once again can't drive to the trailheads. As one who helped make all that improvement possible in the early '90s, it breaks my heart. As a taxpayer, it makes me angry. And as a former Forest Service employee who once took pride in such work, it makes me sad for those still working who are unable to reverse the downward trend."

The Trump budget falls short

The White House budget plan released for Fiscal Year 2020 proposes $2.7 trillion in spending cuts. This includes a proposed 15 percent cut to the Department of Agriculture and a 14 percent reduction in funding for the Interior Department.

American Forests' President and CEO, Jad Daley, issued this statement in response:

"Raging wildfires, pest infestations, and other damage to our forests are diminishing their capacity to slow climate change, filter our water supplies, create forest jobs, and more. We need to provide the proper restoration and management of forests if we want them to continue delivering these benefits. This is absolutely the wrong time to cut funding from the Agriculture and Interior Departments, which lead federal efforts to help to conserve and restore our forests.

"In fairness to the Administration, this budget does contain some positive measures for forest restoration. Most importantly, this budget makes good on implementing the 'Wildfire Funding Fix', which American Forests helped to enact, by providing increased funding for reducing hazardous fuels to reduce wildfire. But this positive step is counterbalanced by massive cuts that will shortchange programs critical for conserving forests from development, urban forestry, research, and partnership programs. This will leave state, community and private forest owners in the lurch and undermine scientific support for all forest activities on public and private lands alike."

In Alaska, Republican Sen. Lisa Murkowski—nominally regarded as more moderate than the rabid rightwingers in her party—is nevertheless working with the Forest Service to eliminate roadless areas on the Tongass National Forest.

Various types of wood have become extinct

Most people would probably be shocked to learn that U.S. forests have been stripped of some beautiful woods that are no longer available, but had once been commonplace in America.

When used for such purposes as making grand pianos, the now-rare woods imparted a rich tone that's dramatically superior to pianos made today. Sonny Stancarone of Sonny's Piano in Port Jefferson, N.Y. restores these beautiful vintage instruments, and finds his customers are thrilled by the exquisite *timbre from the timber* that cannot be otherwise duplicated by modern and often cheaply made pianos.

Veneer is a thin decorative covering of fine wood applied to a coarser wood or other material. Many beautiful woods like sycamore, amboyna, mahogany and violet wood were used on furniture, along with rich hard woods like ebony or macassar. Other woods popular were rosewood, Carpathian Elm Burl, Brazilian jacaranda, zebra wood, palmwood, Harewood, figured satinwood, Tamo, Curly Eucalyptus, Amboyna Burl, Bethlehem Olivewood, Bois de Rose, Birdseye Maple, Bocote Pen, Afrormosia, Ajo, Monkey Puzzle, Podocarp, Ramin, Red Stinkwood, Argentine Lignum, and calamander. Most people these days have never heard of these woods.

An international agreement between governments was formed in 1973, called the Convention on International Trade of Endangered Species (CITES), which lists woods that cannot be traded due to scarcity. One of the most common instances of this is with guitars made of Brazilian Rosewood. In these instances, it is illegal to take such items across international borders without a proper export permit.

Angela Nelson of the Mother Nature Network advises that "More than 6,400 trees are listed as globally threatened on the [International Union of Conservation for Nature] IUCN Red List, according to the Global Trees Campaign. Over 1,100 trees are critically endangered and in need of immediate conservation action. By some estimates, as much as ten percent of the world's trees worldwide are threatened with extinction—and many of those are in our own backyard."

Ms. Nelson notes that "From the California coast to an Arkansas forest, the United States is home to several species of threatened and endangered trees. Their populations have decreased due to disease, insects and pests, development, logging and more."

Angela Nelson has compiled the following list of 11 trees in America which have an uncertain future, and comments on each:

• **Virginia round-leaf birch:** Found only in—you guessed it — Virginia, this critically endangered tree was believed previously to be extinct. However, according to IUCN, it was rediscovered along Cressy Creek in 1975 and is now "found in highly disturbed second-growth forest" along a tiny stretch of the river in Smyth County. Only about 11 individual trees remain.

• **Loulu:** Fewer than 130 individual Loulu trees remain in the Waianae Mountains on the Hawaiian island of O'ahu.

• **Florida yew:** There's only one known tiny population of this critically endangered tree species: a nine-square-mile section of ravines and bluffs along the Apalachicola River in northern Florida. A lack of regeneration, an increasing deer population and declining habitat quality are the main culprits for the dwindling number of plants.

The United States Botanic Garden, which calls the Florida yew one of the rarest trees in the world, says another reason the trees are endangered is because many

The Florida yew has important medicinal value

are on private land, and endangered species laws do not protect endangered plants on private property.

The botanic garden uses the Florida yew to show why conserving these trees is so important. *The bark on this species of yew contains the cancer-fighting compound taxol, which has proven useful in treating breast cancer, ovarian cancer, some kinds of leukemia and certain kidney diseases.*

• **Catalina mahogany:** This small tree is endemic to Catalina Island with only one single population remaining in the wild.

• **Gowen cypress:** Also known as the dwarf cypress and Santa Cruz cypress, fewer than 2,300 individual Gowen cypress trees are found in just five counties in California.

• **Florida torreya:** This is the second Florida-specific tree on the list, as well as the second yew—in fact, this critically endangered tree is also known as the stinking yew because its leaves, when crushed, give off a turpentine odor. These slow-growing trees, which can be 40 feet tall and 20 feet wide, are native to a 40-mile stretch of the Appalachicola River in northern Florida, though they are rarely found in the wild, according to the University of Florida. The Florida torreya, an evergreen conifer tree, has seen a 98 percent decline in population within the last three generations. Fewer than 600 individual trees remain.

• **Fraser fir:** That's right, your favorite type of Christmas tree is officially endangered. And while chopping down millions of these trees each year might seem to be the cause, the problem is actually an insect. Today, this species is found toward the top of the Appalachian Mountains in southwestern Virginia, western North Carolina and eastern Tennessee. Preserving the Fraser fir is critical to the rare animal species that live in those areas and rely on the tree, the Global Tree Campaign says, such as the northern flying squirrel, Weller's salamander, the spruce-fir moss spider, mountain ash and rock gnome lichen.

• **Longleaf pine:** The IUCN lists this species of pine tree native to the Southeast United States as endangered, but they say it may qualify as critically endangered if the time frame for assessing the threat level was expanded. The declining population of this species is mostly due to logging.

"Longleaf pine was heavily exploited since Europeans settled in the Coastal Plains and served a major forest industry in the region. Its wood is used for sawlogs, stage flooring, plywood, pulpwood and produces poles, fence posts, and piling as it makes straight stems largely free of branches when grown in closed stands. Turpentine and other chemicals can be distilled from the chipped wood," the IUCN says.

• **Two California redwoods:** You can't get much more American than California's redwood forests, mentioned in the chorus of Woody Guthrie's famous folk song, *This Land Is Your Land*. But two redwood species—coast redwoods and giant sequoias—are listed as endangered on the IUCN Red List.

Though many of the trees are in protected areas, such as Redwood National Park, the populations continue to decline due to "inadequate regeneration and natural death of (over)mature trees, which are being replaced by other, competing conifers," according to IUCN.

The fast-growing coast redwood is the world's tallest tree

species, according to the Global Trees Campaign, and the oldest one on record is 2,200 years old. And though giant sequoias, which can grow to over 250 feet tall, still number in the tens of thousands, they were logged extensively in the past and their numbers continue to decline today.

• **Maple-leaf oak:** It's name says maple but make no mistake: This is an oak tree with maple-tree-shaped leaves—a rare species that only grows in steep, rocky forests of the Ouachita Mountains in west central Arkansas and southeast Oklahoma. It's listed as endangered on the IUCN Red List due to habitat degradation, with fewer than 600 individuals left in the wild.

Maple-leaf oak trees grow to 40 or 50 feet tall, the Missouri Botanical Garden says, and the leaves on this deciduous tree of the red oak group are yellowish-green.

Seventy-eight species of oak trees are threatened around the world, including 17 in the U.S., according to Botanic Gardens Conservation International.

Probably the tree that is now endangered but which was best known by prior generations is the **American Chestnut** tree. It once comprised a quarter of all eastern hardwoods, with economic and environmental values unmatched by anything in today's forests.

Tom Horton of American Forests—a group that dates back to 1875—writes that "Chestnuts dominated eastern hardwood forests not only in numbers; an estimated three to four billion trees across more than 30 million acres. Known as 'redwoods of the East,' chestnuts grew fast and big, and lived long, reaching 100 feet in height, with diameters exceeding 12 feet, and attaining an average age of two to three centuries.

"Their bold-grained, blondish wood was strong, easily worked, and extremely rot-resistant, used in everything from barn timbers to pianos, split-rail fences to fine furniture (in which it was often veneered with more fashionable woods like mahogany). It was beloved by timbermen for re-sprouting readily from the stump and reaching diameters of two feet or more in little over half a century; an oak on similar soils would take a couple centuries to add as much wood. 'By the time a white oak acorn has made a baseball bat, the chestnut stump has made a railroad tie,' one advocate boasted."

Horton says that "A mature chestnut's sweet, carroty-tasting nuts—as many as 6,000 from a single tree—were nearly a perfect food for both settlers and their livestock, as well as an array of wildlife from turkeys to bears. They are high in fiber, vitamin C, protein, and carbohydrates, and low in fat."

He described what it used to be like: "Once, their creamy June bloom so festooned the eastern hardwood forests that they looked from afar 'like a sea with white combers plowing across its surface,' wrote the naturalist Donald Culross Peattie."

Importantly, Tom Horton says "Their profusion of bloom supported honeybees and other pollinators. And because chestnuts blossom relatively late, their nut crop was never hit by the late frosts that often diminish the mast of oaks and hickories. The Romans ranked chestnuts alongside the olive tree and the grapevine as plants important to civilization."

Then tragidy struck. "That annual exuberance of the American chestnut began fading from the landscape around 1904, when a blight imported on Asian chestnuts began rampaging from Maine to Georgia. By the 1950s destruction was complete," Horton laments.

"Of literally billions of chestnuts growing in the tree's historic range when the blight hit, only dozens of pre-blight survivors struggle on in the wild today," he notes. "Far more numerous are chestnuts that sprout from the roots of felled forest giants, only to die in a decade or two from the deadly fungus that may never go away."

There is reason for hope, however. Horton says "A modest but historic planting of several hundred little chestnuts has completed their first full growing season in the wild on U.S. Forest Service lands in Virginia, North Carolina, and Tennessee. Researchers say they are strong performers, reaching three to seven feet, some flowering at an earlier age than normal."

"But we're excited," says Meghan Jordan of the American Chestnut Foundation (ACF), which supplied the trees. "This means that our goal after 25 years has moved from breeding a chestnut that can survive to working on landscape-level restoration."

❑

A Green New Deal plan must encompass protections for our woodland areas, both public and private, so that reckless exploitation of them for profit does not continue as it has for so long.

8.

Oceans In Peril

I F IT'S RAINING WHERE YOU ARE, THE OCEANS PLAYED A ROLE. IF YOU drove to work, the seas are absorbing the carbon dioxide from your car. If you ordered seafood for lunch, it may have traveled halfway around the world to land on your plate. No matter where you live on earth, what you do affects the oceans—and what happens to the oceans affects you.

"People think the ocean is a place apart," said Peter Neill, head of the World Ocean Observatory. "In fact it's the thing that connects you—through trade, transportation, natural systems, weather patterns and everything we depend on for survival."

Seen from space the earth is covered in a blue mantle. It is a planet on which the continents are dwarfed by the oceans surrounding them and the immensity of the marine realm.

A staggering 80 percent of all the life on earth is to be found hidden beneath the waves and this vast global ocean pulses around our world driving the natural forces which maintain life on our planet.

The oceans provide vital sources of protein, energy, minerals and other products of use the world over. The rolling of the sea across the planet creates over half our oxygen, drives weather systems and natural flows of energy and nutrients around the world, transports water masses many times greater than all the rivers on land combined and keeps the earth habitable.

Without the global ocean there would be no life on earth.

It is gravely worrying, then, that we are damaging the oceans on a scale that is unimaginable to most people.

We now know that human activity can have serious impacts on the vital forces governing our planet. We have fundamentally changed our global climate and are just beginning to understand the consequences of that.

As yet largely unseen, but just as serious, are the impacts we are having on the oceans.

A healthy ocean has diverse ecosystems and robust habitats. The actual state of our oceans is a far cry from this natural norm.

A myriad of human pressures are being exerted both directly and indirectly on ocean ecosystems the world over. Consequently ecosystems are collapsing as marine species are driven towards extinction and ocean habitats are destroyed. Degraded and stripped of their diversity, ocean ecosystems are losing their inherent resilience.

We need to defend our oceans because without them, life on earth cannot exist.

Oceans cover more than two-thirds of the world's surface. In the past 50 years we have learned more about the earth's oceans than in all of preceding human history. But, at the same time we learned more, we lost more.

Things Are Getting Worse

Disturbing news continues to emerge from beneath the waves. Recently, science journals reported that disappearing deep-sea species could trigger an ocean-wide collapse of sea life, that global warming is destroying coral, and that loss of top predators is knocking ocean ecosystems out of whack.

That's just some of the bad news. The oceans are in peril, and therefore, so are we.

The future food security of millions of people is at risk because over-fishing, climate change and pollution are inflicting massive damage on the world's oceans, marine scientists have warned.

The crisis has many dimensions. Runaway residential development destroys coastal wetlands and estuaries. Increased runoff from our farms and cities pollutes streams, rivers and eventually the sea. Commercially valuable fish species—cod, haddock, flounder and salmon off the Atlantic coast, salmon in the Pacific—continue to decline. Invasive species pose a growing threat.

High-impact fishing methods are rapidly emptying the sea, disrupting food chains fine-tuned over millenia. They have essentially eliminated the easy-to-catch dinner table fish, pushing fishing operations farther out into the deep ocean where species are more fragile.

In the past 50 years, the populations of every single species of large wild fish have fallen by 90 percent or more. The sharks, tuna, marlins, swordfish, halibut and grouper that have managed to survive are, on average, one-fifth to one-half the size they were 50 years ago.

Overfishing is depleting fish populations worldwide.

In the deep oceans, where Japanese fleets use fishing lines many kilometers long, they used to catch ten fish per 100 hooks; now they are lucky to catch one.

Heavy nets used by industrial bottom trawlers ravage crucial habitat, ripping up sea grasses and coral forests that are up to 2,000 years old. Water quality grows ever-poorer, fouled by a witch's brew of pollutants running off the land or dumped at sea. Nutrients, mostly fertilizers and sewage, have created at least 146 "dead zones" in the world's oceans, with oxygen levels so low that marine life cannot survive.

A "plastic soup" of waste floating in the Pacific Ocean is growing at an alarming rate and now covers an area twice the size of the continental United States, scientists have said. The vast expanse of debris—in effect the world's largest rubbish dump—is held in place by swirling underwater currents. This drifting "soup" stretches from about 500 nautical miles off the California coast, across the northern Pacific, past Hawaii and almost as far as Japan. Scientists believe that about 100 million tons of flotsam are circulating in the region.

Research by marine biologist Boris Worm projects that all commercial seafood species could collapse within 30 years due to overfishing, loss of habitat, and pollution.

The amount of marine life we extract to feed ourselves is astronomical, and some of our fishing methods—dynamite fishing, bottom trawling, cyanide fishing, and other techniques—cause great damage to current and future fish stocks and to the underwater world in which they thrive. Today, 90 percent of the oceans' top predators are gone. Entire populations of fish, and the communities and economies they support, have collapsed. Seafloors look like war zones. Corals have been bleached white from chemical runoff. Dead zones— vast swaths of ocean that can no longer support life—are spreading throughout the marine realm.

A massive dead zone, some 8,000 square miles across, forms each summer in the Gulf of Mexico as floodwater flushes nitrogen-rich fertilizer from Midwest farms into the Mississippi River.

Two distinguished panels—the U.S. Commission on Ocean Policy and the Pew Oceans Commission—released major reports that found our oceans and coasts in serious trouble. Like the 9/11 Commission, the Ocean Commission proposed detailed recommendations to avert a crisis in homeland security. The homeland crisis in our oceans, however, is getting worse, not better.

One serious problem, says professor Frank Muller-Karger, is that we have upwards of 14 federal agencies charged with ocean issues. These agencies are, in turn, overseen by more than 60 Congressional committees and subcommittees.

The independent Pew Oceans Commission has called for immediate reform of U.S. ocean laws and policies to restore ocean wildlife, protect ocean ecosystems, and preserve the ecological, economic, and social benefits the oceans provide.

Energy platforms may have severe negative impacts on local fisheries. Reports have established a connection between oil and gas rigs and elevated mercury levels and the dangerous ciguatera toxin in surrounding wild fish. Moreover, energy platforms can cause environmental damage, and pose navigational hazards when they spill oil (as will be discussed later in this chapter).

Overfishing

The schools of cod that had greeted the first Europeans in the New World—cod five and six feet long that you could catch by dipping a basket in the sea—were already reduced by the 1950s. But the damage had barely begun.

Pretty soon new technology was at work: fish-finding sonar, big factory ships that could wait offshore for months, helicopters for chasing tuna. The equipment was so good that fishermen could keep bringing in sizable catches right until the moment that the populations

This woman caught a 363-pound fish.

crashed for good. Once Canadian cod fishermen were able to efficiently locate the nurseries where the fish spawned, for instance, they were able to drag their trawls right through them. On paper everything seemed fine until 1992 when, finally, the nets came up empty.

The Canadian government imposed a moratorium on cod fishing that year. The drastic move instantly threw 40,000 people out of work, sparking angry protest and widespread shock. Hundreds of communities were wiped out.

But it was a case of bolting the dock door after the fish had fled. Cod populations have been cut by 99 percent, and the ecology of the ocean may have been changed so profoundly that they're never coming back to the level they once were. The cod stocks in the North Sea and Baltic Sea are now heading the same way and are close to complete collapse.

Sue Bailey of the *Canadian Press* writes that Newfoundland's northern cod are starting to make a comeback after 27 years, but that a Canadian study says that while cod stocks continue to grow, they are still in a "critical zone."

"The union representing fishermen and plant workers, however, wants to immediately expand the relatively small commercial cod fishery, saying it would not stunt the growth... [But] stocks are still well below what would be needed to sustain larger-scale fishing."

The authors of a *Nature* study say that overall, we would need to cut total ocean fish catches by 50 percent to give stocks any chance to recover. Instead, fishing pressure may actually be increasing. As big species are wiped out, the fleets go for smaller fish. Pilchard and anchovy catches are way up, in part so that they can be ground into fish-meal and fed to those farmed salmon you find in the supermarket.

Scientists, politicians, and fishermen alike know there's a problem. Smaller fish and smaller catches suggest that the world's oceans are no longer producing at their full potential. The bounty of the sea is becoming less generous. Improvements in technology have made it easier for fishermen to find and harvest more fish than ever before while demand for sea life products—which are consumed by both the rich and poor—is at an all time high.

Many marine ecologists think that the biggest single threat to marine ecosystems today is overfishing. Our appetite for fish is exceeding the oceans' ecological limits with devastating impacts on marine ecosystems. Scientists are warning that overfishing results in profound changes in our oceans, perhaps changing them forever. Not to mention our dinner plates, which in the future may only feature fish and chips as a rare and expensive delicacy. The fish don't stand a chance.

More often than not, the fishing industry is given access to fish stocks before the impact of their fishing can be assessed, and regulation of the fishing industry is, in any case, woefully inadequate.

The reality of modern fishing is that the industry is dominated by fishing vessels that far outmatch nature's ability to replenish fish.

On the one hand, health officials urge us to eat less red meat and more seafood. But consumer demand for seafood is depleting our ocean supplies.

Giant ships using state-of-the-art fish-finding sonar can pinpoint schools of fish quickly and accurately. The ships are outfitted like giant floating factories—containing fish processing and packing plants, huge freezing systems, and powerful engines to drag enormous fishing gear through the ocean. Put simply: the fish don't stand a chance.

Populations of top predators, a key indicator of ecosystem health, are disappearing at a frightening rate, and 90 percent of the large fish that many of us love to eat, such as tuna, swordfish, marlin, cod, halibut, skate, and flounder have been fished out since large-scale industrial fishing began in the 1950s. The depletion of these top predator species can cause a shift in entire oceans' ecosystems where commercially valuable fish are replaced by smaller, plankton-feeding fish. This century may even see bumper crops of jellyfish replacing the fish consumed by humans.

These changes endanger the structure and functioning of marine ecosystems, and hence threaten the livelihoods of those dependent on the oceans, both now and in the future.

Bycatch is also hurting fish stocks

Many fisheries catch fish other than the ones that they target and in many cases these are simply thrown dead or dying back into the sea. In some trawl fisheries for shrimp, the discard may be 90 percent of the catch. Other fisheries kill seabirds, turtles and dolphins, sometimes in huge numbers.

Estimates vary as to how serious a problem bycatch is. Latest reports suggest that around eight percent of the total global catch is discarded, but previous estimates indicated that around a quarter of might be thrown overboard. Simply no one knows how much of a problem this really is.

The incidental capture, or bycatch, of mammals, seabirds, turtles, sharks and numerous other species is recognized to be a major problem in many parts of the world. This figure includes non-target species as well as targeted fish species that cannot be landed because they are, for instance, undersized. In short, anywhere between 6.8 million and 27 million tons of fish could be being discarded each year, reflecting the huge uncertainties in the data on this important issue.

The scale of this mortality is such that bycatch in some fisheries may affect the structure and function of marine systems at the population, community and ecosystem levels. Bycatch is widely recognized as one of the most serious environmental impacts of modern commercial fisheries.

The Victims

Different types of fishing practices result in different animal/species being killed as bycatch: nets kill dolphins, porpoises and whales, longline fishing kills birds, and bottom trawling devastates marine ecosystems.

It has been estimated that a staggering 100 million sharks and rays are caught and discarded each year. Tuna fisheries, which in the past had high dolphin bycatch levels, are still responsible for the death of many sharks. An estimated 300,000 cetaceans (whales, dolphins and porpoises) also die as bycatch each year, because they are unable to escape when caught in nets.

Birds dive for the bait planted on long fishing lines, swallow it (hook included) and are pulled underwater and drowned. Around 100,000 albatrosses are killed by longline fisheries every year and because of this, many species are facing extinction.

Bottom trawling is a destructive way of "strip-mining" the ocean floor, harvesting the species that live there. As well as the target fish species, this also results in bycatch of commercially unattractive animals like starfish and sponges. A single pass of a trawl removes up to 20 percent of the seafloor fauna and flora. The fisheries with the highest levels of bycatch are shrimp fisheries: over 80 percent of a catch may consist of marine species other than the shrimp being targeted.

Many technical fixes exist to reduce bycatch. Turtle exclusion devices are used in some shrimp fisheries to avoid killing turtle species. In the case of long-line fisheries, the process of setting the hooks can be changed and bird-scaring devices employed which radically cut the numbers of birds killed. To avoid dolphins being caught in nets other devices can be used. Pingers are small sound-emitting and dolphin-deterring devices that are attached to nets, but they are not always effective. Escape hatches (consisting of a widely spaced metal grid, which force the cetacean up and out of the net) have also been employed.

Although these devices may have a role to play, they cannot address the whole problem. Such devices need continual monitoring to check how well they work and assess any potential negative effects they may have. Realistically they will probably only be used in areas with well-developed fishery management and enforcement agencies.

On a global level, probably the only effective way to address the problems of bycatch is to control fishing effort. This will be best achieved through the creation of marine reserves. Nonetheless, in the case of highly mobile species such as seabirds and cetaceans, the only effective way of preventing bycatch is to discontinue the use of particularly damaging fishing methods. However, there is resistance by large segments of the fishing industry to adopting these methods.

Ripping Off the Pacific Communities

Instead of trying to find a long-term solution to these problems, the fishing industry's eyes are turning towards the Pacific.

Pacific people have fished the ocean for thousands of years, managing traditional fishing grounds in a sustainable way. Today a fleet of locally-based vessels, owned by foreign and local companies, catch about 200,000 tons (ten percent of the total catch) of tuna a year. But increasing numbers of industrial distant water fishing boats are moving into the Pacific, taking about 1,800,000 tons (90 percent of the total catch). Instead of reducing their fishing effort and the number of boats when they fish out their own fishing grounds, countries like China, Korea, Taiwan, Japan, the U.S. and the E.U. simply move on to the next fishing ground: the Pacific.

To make matters worse, the practice is also financially exploitative—the economic return from access fees and licences to the region is a mere five percent or less of the $2 billion the fish is worth on the market.

Of course, the returns from pirate fishing are non-existent. Pirate fishing boats do not comply with any rules and only disadvantage the region.

The Pacific is at a crossroads. One path leads to sustainable and equitable fisheries, a healthy marine environment and stable and prosperous island communities. The other path leads to the collapse of the major tuna fishery and loss of livelihood and food supply for the people of the Pacific.

The rapacious quest for seafood

The fundamental problem facing the management of the world's fisheries is the "tragedy of the commons." Since in many areas, ocean fish are essentially an open access resource there is no incentive for an individual fisherman to restrain his catch. If an individual fisherman conserves for tomorrow, he is only providing someone else with the resource today.

As some wealthy countries have depleted their own fisheries, they have moved into poor countries, buying fishing rights—often from corrupt bureaucrats—and then continuing their unsustainable plunder of marine resources.

Some have argued that a global network of marine parks could

Many developing nations—especially in the Caribbean—rely on reef tourism as a crucial part of their economies. Here's a snorkler off the coast of Jupiter, Florida.

help protect and restore some of the world's most impacted fisheries, while aquaculture—which has experienced tremendous growth in recent years—could supplement protein needs. But in order for any sort of conservation plan to be successful, all the involved parties must be brought to the table to hammer out a mutually agreeable plan. If past summits are any indication, this could take a very long time.

Sixty-four percent of ocean areas fall outside national jurisdictions, making it difficult to reach international consensus or to stop illegal fishing—a growing concern as high-tech ships scour the high seas.

Coral Reefs

Besides being enclaves of biodiversity, coral reefs perform other important functions. In Asia, coral reef-based fisheries provide a quarter of the fish that help to feed a billion people. They are also critical mechanisms for protection throughout the earth's tropical regions. Without coral reef barriers, coastal areas will become more vulnerable to the kind of devastation caused by hurricanes like Katrina.

Coral reefs are often described as the "tropical rainforests of the oceans." But marine biologists sometimes use another analogy: that of the canary in the coalmine. These birds were used by miners as an early warning for lethal gas; corals, too, are extraordinarily sensitive to environmental change. For Nancy Knowlton, a scientist at the Smithsonian Natural History Museum, it's an apt description. "If that's the analogy, then the canary has passed out on the floor of the cage. Coral reefs are potentially immortal. They only have to die if we make them."

Brainless, immobile and with only the most primitive nervous systems, coral polyps have built some of the most magnificent structures on our planet. They protect us, feed us and astound us with their beauty. The reefs are home to a quarter of all ocean species during some portion of their lives—and since 1980, about 30 percent have disappeared.

"Its plight is bad news for all of us, but will horrify anyone who has put on flippers, mask and snorkel to experience its magnificence first-hand," comments Rob Sharp. He writes that: "Snorkeling over a reef for the first time, as I did off the coast of Florida, is like floating over a brilliantly colored Garden of Eden... Corals of all shapes and sizes grow in the dappled sunlight. Vast, bulbous species covered with beautifully etched crenulations look like the intricate folds on the surface of the human brain. Others resemble petrified trees, their branches sticking up like fingers, or flat pancakes woven with intricate lacework. Waving sea fans drift back and forth with the gentle pulse of the waves, a hypnotic motion that sends you into a trance—like a state of awe.

"And then there are the fish—lots of fish," Sharp recounts. "Nothing quite prepares you for the variety of sizes, colors and shapes swimming in and out of the coral latticework. There are iridescent blue ones with fins like a quiff. There are green ones with metallic scales, each a slightly different hue from the next, like the scaly armour of a Scythian warrior. A much larger fish with camouflaged skin and a big, ugly head spies me with his swivelling eyes and tries to hide, ostrich-like, behind a skinny staghorn coral; a huge ray, five or six feet long. glides effortlessly past, trailing a menacingly spiked tail in its wake."

This scene is repeated everywhere on earth where tropical reefs form—which is just about anywhere on the vast, watery belt around the equator. The biggest of them all is the Great Barrier Reef of Australia. It took more than 10,000 years to get to where it is today, growing at a rate of a centimeter or two each year.

Corals and plankton are at risk of being destroyed by the rising acidity of the world's oceans as the waters absorb carbon dioxide from the air, scientists have warned. The only solution, they say, is drastic cuts in carbon dioxide emissions, far beyond those called for by the Kyoto treaty.

The most recent data shows that levels of ocean acidification predicted for the not-too-distant future—about a three-fold increase over current levels—have already been measured in some U.S. coastal waters.

Acidic seawater is toxic to eggs and developing fish, and inhibits the ability of corals, certain plankton and other animals to build shells and corrodes them.

The other problem is that warmer water temperatures caused by climate change can result in coral bleaching. When water is too warm, corals will expel the algae (zooxanthellae) living in their tissues causing the coral to turn completely white. In 2005, the U.S. lost half of its coral reefs in the Caribbean in one year due to a massive bleaching event.

Coral reefs, already dying from warming waters, could develop symptoms similar to osteoporosis.

Curiously, not all bleaching events are due to warm water. In January 2010, cold water temperatures in the Florida Keys caused a coral bleaching event that resulted in some coral death. Water temper-

This illustrates coral bleaching that took place in just a 40-day period. In the photo at left, the coral as it normally looks. On the right, after coral bleaching.

If this photo were in color, you could see that all of the water in this picture is solidly a blood red—shockingly so—after dozens of whales were killed by villagers at the Faroe Islands. Irresponsible fishing is endangering all the world's oceans.

atures dropped 12.06 degrees Fahrenheit lower than the typical temperatures observed at this time of year. Researchers are evaluating if this cold-stress event will make corals more susceptible to disease in the same way that warmer waters impact corals.

The bottom line is, the extremes we are experiencing due to climate change are harming coral and other ocean elements.

Irresponsible fishing continues worldwide

As just one shocking example of the mindless exploitation of our seas by people who don't care what harm they are doing to the environment, Cambridge University student Alastair Ward, 22, photographed the traditional "whale driving" (photo above) carried out by the community in the bay in Sandavágu. The Faroe Islands are located in the North Atlantic between Norway and Iceland and are made up of 18 tiny islands.

Pilot whale meat and blubber are a food source that will help feed the 50,000 Faroese through winter. Locals have been carrying out the hunts for centuries, but the gruesome images like the one on this page will shock anyone who cares about responsible fishing.

Ward said he had been stunned by the sheer number of whales in the bay. "They were driving them into the bay, prodding them with their oars. Once they got close enough, the whole town sprinted in and started hacking at them," he shuddered. "Even the children were getting involved, pulling on the ropes and jumping on the carcasses. We just sat there speechless and a bit upset but you couldn't really pull yourself away."

The inhabitants consider whaling a community activity open to everyone, which they say is regulated by national laws and is conducted in a way to cause as little suffering to the whales as possible. About 800 whales are killed each year.

But Mr. Ward said he had been concerned over the methods of the whale killings. He told the BBC "The squealing from the whales was horrible. They were putting hooks on ropes in their blowholes to pull them in and then hacking at them with knives. They didn't die in a very humane way."

And then there is the massive poisoning of our oceans

The oceans face a massive and growing threat from something you encounter everyday: plastics. An estimated 17.6 billion pounds of

A beach littered with discarded plastic bottles and other debris.

plastic leaks into the marine environment from land-based sources every year—this is roughly equivalent to dumping a garbage truck full of plastic into the oceans every minute.

According to *Oceana.org*, "As plastics continue to flood into our oceans, the list of marine species affected by plastic debris expands. Tens of thousands of individual marine organisms have been observed suffering from entanglement or ingestion of plastics permeating the marine environment—from zooplankton and fish, to sea turtles, marine mammals and seabirds.

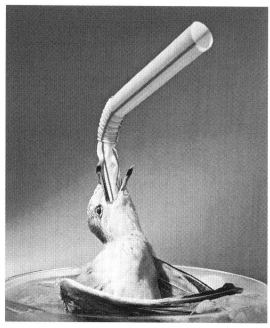

A heartbreaking photo of a bird suffering with a plastic straw stuck in its throat.

"Plastics never go away. Instead, they break down into smaller and smaller pieces, which act as magnets for harmful pollutants. When eaten by fish, some of those chemical-laden microplastics can work their way up the food chain and into the fish we eat.

"Plastics in our oceans threaten the viability of critical marine ecosystems, but marine plastic pollution is not just a problem for our oceans. The extent to which we, too, are being affected by the plastics that have become so ubiquitous in our environment—in our food, water and air—is a topic of extensive research.

"Unfortunately, one of the most popular solutions to plastic pollution falls far short. A meager nine percent of all plastic waste generated has been recycled. Recycling alone is not enough to solve the plastics crisis. To have an impact, we must reduce the amount of single-use plastic being produced at the source. Oceana's plastics campaign will urge companies to adopt alternatives for single-use plastic packaging."

One thing you can do to help prevent this problem is don't use plastic straws when you dine out. An organization called Our Last Straw notes that "Americans use millions of plastic straws a day. Those straws litter our streets, lands, shorelines, and oceans. Plastic drinking straws are among the top ten contributors to marine debris pollution. They do not biodegrade but break down into smaller microplastics that have made their way into our food chain and the deepest trenches of our oceans. The research and statistics on the impacts of plastic straws across the globe are alarming.

"As major distributors of straws, it is imperative the hospitality industry leads the charge for change working to protect our planet and everyone on it."

Our Last Straw is a coalition of restaurants, bars, cafes, hotels, event venues, and organizations across the D.C. metropolitan region and beyond on a mission: Eliminate single-use plastic straws.

The group says that "In all of our efforts, Our Last Straw is mindful that plastic straws may be a necessity for many people with disabilities. All changes in business practices and laws will include accommodations for those requiring plastic straws and will not place greater burdens on people with disabilities who need to use a plastic straw." But as for the rest of us, don't use those plastic straws!

A warning from scientists about the oceans' color change

This next prediction is not only strange, it's ominous. The earth's oceans will turn a deep green by the end of the century as climate change affects the planet, a new 2019 study has predicted. Rising global temperatures will mean that over half of the world's waters will shift color by 2100, according to the report.

This is because warmer waters will change how populations of tiny aquatic creatures called algae grow.

Blue regions such as the subtropics will become even more blue, as algae—also known as phytoplankton—are killed off by the extreme heat.

Green regions such as areas near the poles will turn a deeper green as warmer temperatures trigger huge blooms of the small organisms, according to the study.

As well as turning the world green, the changes could wreak havoc on earth's ecosystems, scientists warned.

"It could be potentially quite serious," warned Dr. Stephanie Dutkiewicz, from the Massachusetts Institute of Technology.

"If climate change shifts one community of phytoplankton to another, that will also change the types of food webs they can support."

Algae reflects green light, and so causes the oceans to appear greener when pictured from space.

Climate experts created a computer model that predicted how the creatures might grow and change the color of our planet in the event of extreme climate change.

It looks at how the sea's color might change if the global temperature would have gone up to 3 degrees celcius by 2100.

The changes were dramatic, with large swathes of the ocean turning green by the end of the century.

"The changes won't appear huge to the naked eye, and the ocean will still look like it has blue regions in the subtropics and greener regions near the equator and poles," said Dr. Dutkiewicz. "But it'll be enough different that it will affect the rest of the food web that phytoplankton supports."

This follows the news that "hot and dry" years have doubled since 1931 due to severe global warming.

What You Can Do

We're all participating in the oceans' demise. We can also be part of the solution by making informed choices in restaurants or at the fish counter. We don't have to eat swordfish or tuna, the apex predators that are the lions of the sea. Only about one percent of tuna comes to the market to be sold fresh. The rest goes to the cannery, because canned tuna is America's most popular fish.

Our survival is not dependent on eating orange roughy that lives 1,500 feet down and is devastated by fishing because it doesn't reproduce until it's 30 years old.

If you need inspiration, take your child or grandchild to the aquarium or the beach. Vote for candidates that understand we're facing an environmental crisis—individuals who will fight to put the health of our planet above corporate profits.

"We have forgotten what we used to have," observes Jeremy Jackson of the Scripps Institution of Oceanography. "We had oceans full of heroic fish—literally sea monsters. People used to harpoon ten-foot-long swordfish in rowboats. Hemingway's *Old Man and the Sea* was for real."

So were passenger pigeons darkening the sky; so were buffalo herds shaking the plains; so were ancient forests piercing the sky. Now there are only echoes—and even those we hardly can hear. Republicans still salivate at drilling for oil in the Arctic National Wildlife Refuge, breeding grounds of one of the world's last big caribou herds. Maybe it's a good thing our memories are so short. Otherwise, we couldn't live with ourselves.

9.
The Dangerous Government and Media Cover-up To Protect Big Oil

THE IMAGES WERE HORRIFYING AND DEPRESSING: CRUDE OIL gushing biliously like smoke from a manmade hole in the earth a mile beneath the surface of the Gulf of Mexico. Thousands of endangered sea animals and birds coated in oil, dying or already dead. Tar sludge washing up on what were once pristine beaches. Fragile marshes rapidly being suffocated by the black viscous liquid. The livelihoods of fishermen and oyster-shuckers being destroyed along 1,300 miles of the Gulf's coastline.

For months, this disturbing tableau was displayed daily across our television screens. Then apologists for British Petroleum (BP)—whose Deepwater Horizon oil rig exploded on April 20, 2010, killing 11 workers and spewing what official estimates claimed were 205.8 million gallons of crude oil uncontrollably into the Gulf of Mexico—tried to minimize the disaster. The official story that emerged from BP and the Barack Obama administration—and still echoed by most of the national media—is: the oil was soon gone; the danger quickly passed and was exaggerated; the dispersants were effective in keeping oil from reaching the shore; the oil that did reach shore was mostly weathered and not toxic; and federal officials found no unsafe levels of oil in air, water or food samples and no evidence of illness due to oil or dispersant use.

The federal government even issued a far-fetched report within days of the oil rig being capped asserting that most of the oil had been captured, skimmed or dispersed naturally or by chemicals. The Associated Press claimed that "a thriving microbial ecosystem has developed to consume the oil," and quoted officials as saying that tiny organisms in the ocean had quickly broken up 84 million gallons of crude—a claim which many scientists still find preposterous. Yet the feds also said "only" 53 million gallons of oil remain unaccounted for.

The Business Insider reported at the time, the "damage has all been vastly overblown." It said "the government thinks the oil spill is not a significant health risk for humans. Sure there are plenty of people reporting headaches and other purportedly oil-related ailments, but these could be temporary or imagined. Remember, oil leaks do occur naturally."

Time magazine ran an article under the byline of Michael Grunwald, who insisted that while "shoreline teams have documented more than 600 miles of oiled beaches and marshes...the beaches are fairly easy to clean, and the beleaguered marshes don't seem to be suffering much additional damage." Grunwald credited radio talk show host and the demagogic Rush Limbaugh with being a voice of reason in minimizing the effects of the spill, and quoted Ivor Van Heerden, (who worked for an oil spill response contractor, was being paid by BP money, and appeared in a BP video), as saying "there's a lot of hype, but no evidence to justify it." *Time* described the spilled oil as "unusually light and degradable," said it was "dissolving quickly," and concluded "anti-oil politicians, anti-Obama politicians and underfunded green groups all have obvious incentives to accentuate the negative in the Gulf."

The unctuous attempts by *Time* and other big media to spin this story in favor of BP, (so as to protect a major source of advertising revenue), are exceeded only by the disingenuous statements of government agencies and officials of British Petroleum. What we actually witnessed for months after the disastrous oil spill was a concerted disinformation campaign designed to protect the profits—first and foremost—of a giant oil company and its investors.

The real-life experiences of the people along the Gulf Coast tell a far different tale. Many of them have not only had their livelihoods destroyed, but their lives as well, and problems are likely to linger for a generation.

The spilled oil covered 44,000 square miles—an area about the size of the state of Kansas. It was conservatively estimated to have cost the tourism industry $22.7 billion; the $2.5 billion fishing industry was also devastated. *Bloomberg* reported the spill undermined consumer confidence—understandably so. The U.S. Treasury lost at least $53 billion in uncollected drilling fees from the oil giants, and BP was claiming a $10 billion write-off on its taxes for costs incurred before the spill was even cleaned up.

The unprecedented and unwise use by British Petroleum of highly volatile chemical "dispersants" to break up the oil on the surface

The British Petroleum (BP) Horizon oil well explosion and spill was one of the world's biggest catastrophes—and was preventable.

of the water, "created an eco-toxioclogical experiment," according to Ron Kendall, a toxicologist from Texas Tech University. He told a Senate hearing that "The bottom line is that a lot of oil is still at sea dispersed in the water column [and] it's a big ecological question as to how this will ultimately unfold."

Vast oil plumes—a toxic mixture of dispersants and methane—continue to lurk below for years, expanding the eutrophic dead zones, killing deep-water corals and giant squid, and wiping out vast populations of plankton, fish, and larger marine animals for years to come. In the years since, this poisonous stew has drifted to other areas of the ocean, causing a loss of species and biodiversity, and tainting the seafood we eat.

The damage is not confined to the coastal region. Those Americans living thousands of miles away from the spill can take no comfort in not being close to the disaster, as it has affected us in many ways. Since 65 percent of the grain our farmers export—corn, soybeans and wheat—travels on barges down the Mississippi and through the Gulf, cargo ships needed to have oil washed off their hulls before moving upriver. Some farmers opted to ship via more expensive truck or rail to other ports.

There was also a health risk to other areas of the country. The millions of gallons of dispersants which were dumped in Gulf waters evaporated, and were carried north on air currents, and deposited on the central plains as rain.

Oil has also affected the birds we see in other parts of the nation. The Gulf coast wetlands are the equivalent of a busy airport hub, with 110 species of migratory songbirds and 75 percent of all migratory waterfowl making a stopover there.

The oil spill fouled the plans of many of the so-called "snowbirds" from northern climes who vacation in sunbelt coastal communities during the winter and had to change their travel plans.

Even the clean-up efforts inflicted more damage on the area. The 5,600 vessels that took part in the oil spill operation in the Gulf constitute the largest fleet assembled since the Allied invasion of Normandy. BP also hired truckloads of inexperienced oil spill responders—shrimpers, unemployed workers, college students, migrants —who then trampled, drove over and even ran airboats across fragile habitat in their poorly coordinated attempts to deal with the petroleum.

While the Obama administration issued an order to BP restricting the use of controversial dispersants like Corexit to "rare cases," the Coast Guard ignored this directive, giving British Petroleum every waiver it requested—dozens of them. At least 1.8 million gallons of highly toxic dispersants were used—chemicals which BP is not allowed to use in England for health and safety reasons, but felt free to dump in our waters. Airplanes sprayed the chemicals across the Gulf, like Agent Orange in Vietnam, and the company even injected them 5,000 feet below the surface of the water, at the well head.

EPA whistleblower Hugh Kaufman explains: "Corexit is ...used to atomize the oil and force it down the water column so that it's invisible to the eye. In this case, *these dispersants were used in massive quantities...to hide the magnitude of the spill and save BP money.*" The result was that more oil sank out of sight and thus was out of reach of the cleanup operation. But its affects on sea life were devastating. The EPA admits it does not know what the long-term ramifications of Corexit will be.

Susan Shaw, director of the Marine Environmental Research Institute, insists that "Corexit dispersants, in combination with crude oil, pose grave health risks to marine life and human health and threaten to deplete critical niches in the Gulf food web that may never recover."

Unfortunately, we can't always rely on the federal government to be honest with us. (Remember, the feds declared the air at Ground

Zero safe following the terrorist attacks on 9/11, just as they had vouched that Agent Orange was harmless several decades earlier).

Under fire for having allowed the Coast Guard to issue waivers to BP to continue using the toxic dispersant, the EPA did a very limited test and announced that the combination of oil and dispersants is no more toxic to sea life than oil alone. But the agency only conducted tests on two forms of sea animals—one type of shrimp and one small fish, and only tested to see how much was necessary to kill them. Susan Shaw was among those scientists who scoffed at the tests, noting "There are at least 15,000 marine species in the Gulf that could be impacted by the dispersed oil." And she pointed out that the EPA completely failed to address the concern that the properties that facilitate the movement of dispersants through oil also make it easier for them to move through cell walls, skin barriers and membranes that protect vital organs, underlying layers of skin, the surfaces of eyes, mouths and other structures. Scientists have found signs of an oil-and-crude mix in larvae that is consumed by various sea animals, a clear indication that the dispersants in the BP oil spill broke up the oil into toxic droplets so tiny that they could easily enter the food chain. Fish, shrimp, crab and bluefin tuna are at special risk.

Even at beaches along the Gulf that seemed clean, where families with children were allowed to frolic in the water, there were health

This poor, suffering creature—one of many—was covered in oil, an unspeakable tragedy because of greedy oil interests and politicos who let this happen.

risks. When the WKRG-TV news team in Mobile took samples of water and sand from Alabama beaches and had them tested by a chemist, the results showed concentrations of oil as high as 221 parts per million (ppm), when five ppm should be the maximum. One sample exploded in the lab before it could be tested.

There was also evidence of dangerous levels of oil in the air. A preliminary study commissioned by Guardians of the Gulf, a community-based nonprofit organization in Orange Beach, Alabama, found that nightly air inversions—common in the area during the summer and fall—were trapping pollutants near the ground. Total Volatile Organic Compounds (VOCs)—including the carcinogen benzene, and oil vapors—reached 85 to 108 ppm. (For comparison, the federal standard for 15-minute exposure to benzene is five ppm.) As long ago as 1948, the oil industry itself admitted that "the only absolutely safe concentration of benzene is zero."

Such high levels could explain the bout of respiratory problems, dizziness, nausea, sore throats, headaches, and ear bleeds that were being reported by residents along the Gulf.

A new study says that oil spills can also boost levels of arsenic in seawater, and because the substance is accumulative, it becomes more concentrated and lethal the more it moves up the food chain. This can create a toxic ticking time bomb, which could threaten the fabric of the marine ecosystem in the future and pose health risks to anyone who eats food from the affected area, whether it be from on land, or sea. Petroleum contamination is known to cause cancer and brain damage, but the decision to declare the food safe or not is fraught with political implications, and influenced by everyone who has a stake in the outcome.

The FDA told Congress it had "implemented a surveillance sampling program of seafood products at Gulf Coast area primary processing plants" to "provide verification that seafood being harvested is safe to eat." But according to AOL News, most officials and business owners in the Gulf said they hadn't seen FDA personnel on the shrimp or crabbing boats, on the docks or in the processing plants. Harriet Perry, director of the Center for Fisheries Research and Development at the Gulf Coast Research Laboratory, warned "The oil's going to get into the food chain in a lot of ways."

A Massive Cover-up

From the outset, federal agencies aided and abetted BP's lies and cover-ups. They claimed the spill was no more than 1,000 barrels

A farcial photo op: President Obama looks at a pea-sized tar ball rather than ones larger than basketballs that were washing up on other despoiled Gulf beaches.

a day when it turned out to be at least 53,000 barrels, or a total of 170 million gallons; restricted media access, and threatened journalists with felony charges and $40,000 fines. Uniformed police officers in Louisiana harassed photographers at public beaches, and federal troops were guarding BP headquarters—proof positive that our government protects the oil giants. BP threatened workers with firing if they talked to anyone about anything. The company tried to hire prominent scientists and environmentalists and silence them from speaking out. They bought up all the Google ad searches. They spent $50 million on a PR campaign instead of using the money to compensate victims. BP resisted entreaties from scientists that they be allowed to use sophisticated instruments at the ocean floor that would give a far more accurate picture of how much oil was really gushing from the well.

Sources within FEMA and the Army Corps of Engineers claimed the Obama White House resisted releasing any "damaging information" about the oil disaster. In what BP insiders privately derided as a "pony and balloon show," President Obama visited beaches where damage was minimal and there were pea-sized tar balls, not the areas besmirched with thick, black crude the size of boulders. Even Obama's much-vaunted agreement with BP that the company would pay $20 billion in damages was a sham, and let his biggest financial contributor off easy. The U.S. military has quietly continued to carry on a major, multi-billion dollar business partnership with BP, despite the company's disastrous environmental record.

Then it turned out that even Kenneth Fineberg, appointed by the President to be a "claims czar" was actually on the payroll of British Petroleum and he refused to disclose how much he and his lawyers were being paid. The notion that he was independent had been relentlessly promoted by the Obama administration, BP, the media, and Feinberg himself, who traveled up and down the Gulf Coast telling victims of the oil blowout that they would be "crazy" to sue BP—his employer—assuring them they could do much better through his claims facility.

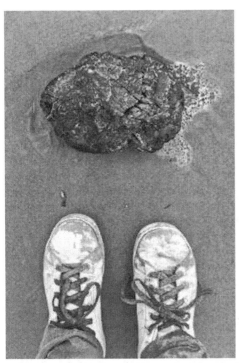

Tar balls the size of basketballs were washing up on beaches. But BP kept the news media as far away as possible.

But when they accepted money from the oil giant, they waived their rights to file future lawsuits. Feinberg all but ruled out paying fishermen who operated on a cash-only basis. Fishermen engaged in cleanup saw their earnings subtracted from any settlement they eventually received. Those suffering mental health problems received no compensation— "You have to draw the line somewhere," Feinberg said. Those who emerged with chronic ailments later, but have already accepted a payout from Feinberg, will in many cases have received no further compensation.

What's At Stake

In the aftermath of the BP oil spill, most Americans were shocked to learn that for decades, our government has been greenlighting risky drilling in the Gulf of Mexico. There are 3,600 active offshore oil and gas platforms in the Gulf alone, and the Associated Press reports there are another 27,000 abandoned wells—600 owned by

British Petroleum—creating "an environmental minefield that has been ignored for decades." There are also tens of thousands of miles of pipeline in the central and western Gulf of Mexico, where 90 percent of the country's offshore drilling takes place.

"There have been thousands of spills from 1990 to 2009," said Walter Hang, head of Toxics Targeting, an Ithaca, NY, company that tracks and analyzes federal hazardous spill reports.

Spills increased dramatically under the Bush and Obama administrations. The federal Minerals and Management Service (MMS) has recorded over 330 significant spills—those over 2,100 gallons—since 1964. At least 324 spills involving offshore drilling have occurred in the Gulf alone, releasing more than 550,000 barrels of oil. Four of these spills even involved earlier equipment failures and accidents on the Deepwater Horizon rig.

The Gulf of Mexico is already in a precarious state, routinely having to absorb environmental insults: overfishing, trawlers raking sea floors, frequent hurricanes. *The New York Times* reported that "runoff and waste from cornfields, sewage plants, golf courses and oil-stained parking lots drain into the Mississippi River from vast swaths of the United States, and then flow down to the Gulf, creating a zone of lifeless water the size of Lake Ontario just off the coast of Louisiana."

Upstream pollution, from 1.5 million tons of nitrogen fertilizer running off Midwest farms and down the Mississippi, has created an area of dead water called a hypoxic zone in the Gulf that is second in size only to a similar zone in the Baltic Sea. That heavy dose of nitrogen every summer encourages algae to grow, which results in a huge feast for bacteria that use up oxygen there, leaving little for fish or anything else.

In addition, the Gulf's floor is littered with bombs, chemical weapons and other ordinance dumped in the middle of the last century, even in areas busy with drilling, and miles outside of designated dumping zones, according to experts who work on deepwater hazard surveys.

Oceanographer Sylvia Earle told a Congressional panel what's at stake: "The Gulf of Mexico is not, as some believe, an industrial wasteland, valuable primarily as a source of petrochemicals and a few species of ocean wildlife that humans exploit for food, commodities, and recreational fishing," she explained. "These are assets worth protecting as if our lives depend on them, because in no small measure, they do."

Earle described the Gulf of Mexico as "a living laboratory,

America's Mediterranean, a tri-national treasure better known for yielding hurricanes, petrochemicals, shrimp and, in recent years, notorious 'dead zones,' than for its vital role in generating oxygen, taking and holding carbon, distributing nutrients, stabilizing temperature, yielding freshwater to the skies that returns as rain and contributing to the ocean's planetary role as earth's life support system."

Worldwide, oil companies have ruined large areas of the Niger Delta, Ecuador and other parts of the world. The ruthless pursuit of corporate profits is such that 80 percent of rivers worldwide no longer support life; 94 percent of the large fish in the oceans are gone; and 120 species per day becomes extinct.

Our knowledge about the nature of the undersea world is still primitive, partly because the methods used for exploring the ocean are still primitive. Surprisingly, no one has yet descended to the greatest depth in the Gulf. We know more about the face of the moon than the bottom of the sea, and are better equipped to live and work in space than we are to explore the oceans on this planet.

What it was like during the oil spill

Out on the water at the time of the BP disaster, the spill started as a rainbow shimmer, then turned to wide orangey-red streamers of oil whipping through the waves as far as the eyes could see. From the air, the Gulf appeared to be bleeding. There were the dolphins hopscotching through the thick gunky oil, struggling sperm whales trailing wakes a mile long. Local residents reported seeing strange sights, such as schools of fish in "boils" on the surface of the water, disoriented, their mouths open, swimming sideways, and upside-down, and running into the side of boats.

Sharks were showing up in surprisingly shallow water just off the Florida coast. Mullets, crabs, rays and small fish congregated by the thousands off an Alabama pier. Fish and other wildlife were fleeing the oil out in the Gulf and clustering in cleaner waters along the coast where, unfortunately, the oxygen levels are being depleted.

Methane concentrations in some areas were 10,000 times higher than normal. John Kessler, an oceanographer at Texas A&M University commented at the time: "This is the most vigorous methane eruption in modern human history." This is likely to cause dead zones, putting further pressures on the marine animals.

As author Naomi Klein journalled, after surveying the scene by air, "the swirling shapes the oil made in the ocean waves looked remarkably like cave drawings: a feathery lung gasping for air, eyes star-

A school of dolphins swimming between two streams of BP oil.

ing upwards, a prehistoric bird. Messages from the deep."

When the weather was calm and the sea was placid, ships deploying fireproof booms attempted to corral the black oil, the coated seaweed and whatever fish or other sea life may have been caught in it, and torch it into hundred-foot flames, sending plumes of smoke skyward in ebony mushrooms. But this was an exercise in futility: it would have taken thousands of such fires just to burn the surface oil, and this pollutes the air with toxins. The National Oceanic and Atmospheric Administration (NOAA), said air in the Gulf was "polluted with organics from the spill," including some constituents of crude oil like benzene and toluene that were "well above maximum concentrations measured" over Los Angeles on a bad day.

The oil began washing ashore where the Mississippi River hits the coast. Louisiana is home to 40 percent of the United States' wetlands and the oil continued to threaten over 400 species of animals. The slick endangered migrating birds, nesting pelicans and even river otters and mink along the fragile islands and barrier marshes.

Driftwood and seashells glazed with rust-colored tar lined the surf along some of the Gulf Coast's formerly clean beaches. A vast stretch of dark, chocolatey syrup left behind at low tide was punctuated by thick patches of crude bubbling on the sand. Dead jellyfish were everywhere. A seagull stood so still it looked like it was made out of a slab of chocolate, another frantically flapped its spread wings to try to shake off the oil, and then another manically pecked at the spots on its chest. When the pelicans and other birds like the northern gannet were not rescued and cleaned quickly, they died quickly because they couldn't fly.

BP and Republicans primarily responsible...of course

This remains a huge scandal, to be sure, as the Deepwater Horizon disaster had the all-too-familiar elements of deregulation, deception, and destruction that characterize the relations between governments and multinational corporations. Like the global financial crisis, it all started with the explosion of a bubble, this time of methane gas.

British Petroleum bragged about being at the forefront of technology. Goldman Sachs and the other behemoths of the financial world also claimed to be at the cutting edge of financial innovation. They all lied, hid information, and speculated behind a facade of corporate imagery built through their deceptive advertising campaigns.

Just like the derivatives that took junk assets into every balance sheet of financial institutions, the Deepwater Horizon disaster has no boundaries. First and foremost, the mounting evidence clearly demonstrates that this tragedy was preventable and the direct result of BP's reckless decisions and actions. Internal memoranda show BP execs felt it would be cheaper to pay off the families of dead workers than operate safely. Even after one blast, when BP plead guilty to a felony, and paid $50 million to settle a criminal investigation, and $21 million for violating federal safety laws, this was regarded by company officers as merely the cost of doing business.

According to *The New York Times*, a confidential survey of workers on the Deepwater Horizon in the weeks before the oil rig exploded showed that many of them were concerned about safety practices and feared reprisals if they reported mistakes or other problems. In the survey, commissioned by the rig's owner, Transocean, workers said that company plans were not carried out properly and that they "often saw unsafe behaviors on the rig."

Some workers also voiced concerns about poor equipment reliability, "which they believed was as a result of drilling priorities taking precedence over planned maintenance," according to the survey.

Vital alarm systems on the oil rig, as well as key safety mechanisms, had been turned off by BP subcontractor Transocean at least a year before the fatal explosion.

The Deepwater Horizon floating oil rig before disaster struck.

In its spill plan, BP claimed it could contain any possible spill by vacuuming up over 20 million gallons of oil per day. That was a bald-faced lie. BP's actual recovery rate after the Deepwater Horizon explosion turned out to be a pathetic two percent of that.

But others besides BP are to blame for the oil spill too. Louisiana, Mississippi, Alabama and Florida have been "red states" for a long time, voting for Republican candidates who for years have been waging a relentless campaign for offshore drilling without limit, ("Drill, baby, drill!"), massive deregulation and complete contempt for environmental oversight, including laws designed to protect workers.

Incredibly, even in the midst of this catastrophe, Republicans opposed delaying new drilling projects for six months, and did so under the guise of protecting jobs (even though this portends irreversible damage to the environment). A favorite GOP talking point was that 30 percent of the oil we use comes from the Gulf. But the nonpartisan Energy Information Administration, says the correct figure was only eight percent.

While criticizing the federal response, the five Republican Governors of the Gulf coast states inexplicably failed to make use of the full contingent of 17,500 highly trained and well-equipped National Guard troops that had been made available to them by the Pentagon to help with the oil spill.

Mississippi Governor Haley Barbour, former head of the Republican National Committee, who had close ties to the oil and gas industry, said audaciously "this oil isn't anything to worry about; it's like the sheen coming off the back of a ski boat. You know, you wouldn't want to wash your face in it, but you don't mind skiing through it." Barbour claimed birds shouldn't have been at risk because the oil coming in was supposedly weathered and non-toxic. At one point, Barbour said oil waste products were akin to "toothpaste." *Toothpaste!*

"Once it gets to this stage, it's not poisonous," Barbour said. "But if a small animal got coated enough with it, it could smother it. But if you got enough toothpaste on you, you couldn't breathe."

Barbour also likened the waste product from the oil spill to an "emulsified, caramel-colored mousse, like the food mousse."

Minnesota Senator Michele Bachmann labeled the $20 billion fund for Gulf victims a "redistribution-of-wealth fund." One hundred-plus largely Republican House members described the fund as a "Chicago-style shakedown" of the polluting oil giant BP.

This was reprehensible given British Petroleum's egregious conduct. BP may have even lobbied for the release of a known terrorist convicted of the 1988 bombing of Pan Am Flight 103 that killed 270 people, including 189 Americans. BP allegedly did so to secure a $900 million oil contract with the Libyan government.

Two Oil Men in the White House

Here we go again. More bad stuff about Bush and Cheney. Between January and March of 2001, incoming Vice President Dick Cheney conducted secret meetings with over 100 oil industry officials allowing them to draft a "wish list" of industry demands to be implemented by the oil-friendly administration. Cheney also used that time to re-staff the Minerals Management Service (MMS) with oil industry flunkies including a cabal of his Wyoming cronies. In 2003, the newly reconstituted MMS deferred to the oil cartel by recommending the removal of the proposed requirement for acoustic switches. The MMS's 2003 study concluded that "acoustic systems are not recommended because they tend to be very costly." (The acoustic trigger costs about $500,000 and could have prevented the Deepwater Horizon explosion).

In 2008, as the price for a gallon of gasoline approached $5, President George W. Bush, at the urging of Mr. Cheney, lifted the executive order banning offshore drilling, that had been implemented by Bush's father following the Exxon Valdez oil tanker spill that occurred off the Alaskan shore in 1989. The House of Representatives agreed to let the 26-year-old moratorium on offshore drilling expire.

A new policy instituted under Bush forsook environmental analysis and fast-tracked permits. Declaring that oil companies themselves were "in the best position to determine the environmental effects" of drilling, the new rules pre-qualified deep-sea drillers to receive a "categorial exclusion"—an exemption from review. The Bushites also put an optimistic spin on things, declaring that a "big spill" would be no more than 1,500 barrels.

Unfortunately, the MMS under Bush devolved into a criminal enterprise. As an investigation by Interior's Inspector General revealed, MMS staff members "frequently consumed alcohol at industry functions, had used cocaine and marijuana and had sexual relationships with oil and gas company representatives." Three reports by the Inspector General describe an open bazaar of illegal gifts, bribes and kickbacks. Industry lobbyists also treated MMS employees to regular golf, ski, and paintball outings, trips to rock concerts and professional sports events. All this was spiced with scenes of female employees providing sexual favors to oil industry execs. But it was really the American taxpayers who were getting screwed.

MMS managers were given cash bonuses for granting risky offshore leases, auditors were ordered not to investigate shady deals, and safety staffers allegedly even allowed oil companies to fill in their own inspection reports in pencil before tracing over them in pen.

There was also the Bush administration's plan to open 300 million acres—in Alaska, the Gulf, and along the east and west coasts—to offshore drilling, that had been published in the *Federal Register* literally at midnight on the day that Bush left office.

But President Obama was to blame too

There's no doubt about it, the Gulf Oil spill was Barack Obama's Katrina. During his campaign for the White House, Mr. Obama opposed offshore drilling, saying correctly that it would "not make a real dent in current gas prices or meet the long-term challenge of energy independence."

The new President was well aware that something needed to be done to stop the abuses at MMS, and in fact he claimed he nominated former Republican Senator Ken Salazar, his "great" and "dear" friend, to do just that. He promised that Mr. Salazar would change the specter of Interior being merely "an appendage of commercial interests."

Shortly after assuming office as Interior Secretary in January 2009, Ken Salazar went to MMS headquarters in Denver and admonished them for what he called their "blatant and criminal conflicts of interest and self-dealing" that had "set one of the worst examples of corruption and abuse in government." Pledging to "set the standard for reform," Salazar declared, "The American people will know the Minerals Management Service as a defender of the taxpayer. You are the ones who will make special interests play by the rules." Dressed in his trademark Stetson and bolo tie, Salazar boldly proclaimed, "There's a new sheriff in town."

But despite all his bravado, Secretary Salazar delayed cleaning house and getting rid of oil industry toadies at Interior's Minerals Management Service, whom he allowed to approve 480 drilling projects with no environmental review, using industry-friendly regulations drafted during the Bush years. (Interestingly, in 2019, he is one of Colorado's highest-ranking lawyers for the oil and gas industry).

Just two months after Barack Obama took office, the application that British Petroleum submitted for its Deepwater Horizon well, was approved. Across the board, the Interior Dept. and MMS under Ken Salazar failed to enforce a plethora of environmental laws, including the Clean Water Act. This despite the fact that BP is the last company that should have ever been granted a permit to drill

Interior Secretary Ken Salazar

for oil in deep water. In only three years alone, it received 760 citations for "egregious and willful" safety violations, described as being those "committed with plain indifference to or intentional disregard for employee safety and health." The rest of the oil industry combined had only received one such citation during the same period.

The initial exploration plan that BP submitted to the federal government reads like a Greek tragedy about human hubris. The company claimed that a spill was "unlikely," and that it anticipated "no adverse impacts" to fisheries, wildlife, beaches or wetlands even if a spill should occur. It insisted that, thanks to "proven equipment and technology," adverse affects would be minimal. Presenting nature as a predictable and agreeable junior partner (or perhaps subcontractor), the report cheerfully explained that should a spill occur, "Currents and microbial degradation would remove the oil from the water column or dilute the constituents to background levels." The effects on fish, meanwhile, "would likely be sublethal" because of "the capability of adult fish and shellfish to avoid a spill [and] to metabolize hydrocarbons."

Even more absurd was BP's assurance that if a spill occurred, there is "little risk of contact or impact to the coastline" because of the company's projected speedy response and "due to the distance [of the rig] to the shore"—about 48 miles.

The paperwork it submitted to the federal government was so sloppy—and obviously unscrutinized—that it contained numerous errors, even listing a website for a Japanese home shopping network as the link to one of its "primary equipment providers for BP in the Gulf of Mexico Region [for] rapid deployment of spill response resources on a 24 hour, seven days a week basis"; and assuring its clean-up plans would protect walruses, sea otters and sea lions—even though no such animals are found in the Gulf.

More seriously, the plan did not contain information about tracking sub-surface oil plumes from deepwater blowouts or preventing disease (viruses, bacteria, etc.) transmission to captured animals in rehab facilities, which was found to be a very serious risk following the Exxon Valdez spill. It also lacked any oceanographic or meteorological information, despite the clear relevance of this data to spill response.

Congressional testimony later revealed that while BP and its competitors had spent $39 billion to explore for new oil and gas, the average investment in research and development for safety, accident prevention and spill response was a paltry $20 million a year.

Shortly before the oil spill, in what he claimed was his hope of attracting Republican support for climate change legislation in Congress, on March 31, 2010 Obama held a press conference with Sec. Salazar and announced he was abandoning a decades-long moratorium on offshore oil drilling along the East Coast from Delaware to Florida, in the Gulf of Mexico and Alaska. In lifting the ban, Obama took oil industry engineers and geologists at their word when they assured him that new technologies and drilling methods had rendered deepwater drilling virtually foolproof. In so doing, Obama ignored his own experts on ocean science.

He claimed he had studied the drilling plan for more than a year and assured that "this is not a decision that I've made lightly." In further remarks two days later, Obama asserted that "oil rigs today generally don't cause spills. They are technologically very advanced." Even that wasn't enough for Sarah Palin—the daft Vice Presidential running mate of Sen. John McCain—who sneered at the Obama administration's slowness to implement its plan: "'Let's drill, baby, drill,' not 'stall, baby, stall!'" she urged.

Only 18 days later, on the eve of Earth Day, BP's 30-story tall, half billion dollar, 32,000-ton Deepwater Horizon rig, exploded like a bomb, killing 11 workers and sinking to the bottom of the sea.

Although government scientists were concerned from the start about the real amount of oil that was gushing forth from the damaged well, the Coast Guard relied on ludicrously low estimates from BP. Barack Obama calibrated his response to the Gulf spill based on flawed and misleading estimates from British Petroleum—and then deployed his top aides to lowball the flow rate at an absurd 5,000 barrels a day. Since he had promised his wife another vacation, two days after the spill—when he should have been deploying a full federal response— Barack took off with Michelle for a long weekend at a golf resort and spa.

It had been clear from the start that his administration had placed a higher priority on interests other than the fate of the Gulf. Two months were to pass before the President even addressed the nation from the Oval Office. It took 70 days before Obama ordered skimmers into the Gulf. He audaciously allowed BP to restrict media access to the area in a violation of First Amendment rights.

Apparently to placate organized labor, Mr. Obama refused to waive the Jones Act, which restricts foreign ships from operating in U.S. coastal waters (thus preserving jobs for unions). His EPA was unwilling to grant a temporary waiver of its regulations that would have permitted various skimmers and tankers (some of them very large) to eliminate most of the oil from seawater, discharging the mostly clean water while storing the oil onboard. Obama allowed the Coast Guard, EPA and Army Corps of Engineers to repeatedly interfere with local clean-up efforts and even take time off in the midst of the crisis.

As the single largest recipient of campaign money from British Petroleum of any candidate, Republican or Democrat, Barack Obama made a promise to the Queen of England and the new Prime Minister, David Cameron, that he'd do everything he could to assure that BP wouldn't go under as it is one of England's major corporations if not its number one corporation. But most Americans surveyed felt BP should pay through the nose to the people of this nation, and if they went out of business, so be it.

Oddly, President Obama deferred to BP to handle the oil spill, despite the oil giant's abysmal safety record, criminal recklessness, preoccupation with its own narrow self-interests, indifference to how its actions impact people and the environment, and ongoing mendacity.

Under existing law, BP was liable for every barrel of oil that was spilled. So they had a powerful incentive to try to cover up how much oil was really flowing from their broken pipe. That's why they used dispersants to try to hide the amount of oil and also why they refused to allow other vessels to enter the area to help with clean-up.

With BP in charge, over 1.58 million gallons of the dispersant Corexit were sprayed into the Gulf. The active ingredient in Corexit is a neurotoxin pesticide that is so lethal to both human and aquatic life, it's been banned for years by the European Union. Corexit is also less effective and more toxic than 12 other brands on the market, according to tests by the Environmental Protection Agency. After hundreds of reported illnesses, the EPA instructed BP to cease using Corexit. But BP did not comply.

Most troubling of all, the government has allowed BP to continue deep-sea production at its Atlantis rig—one of the world's largest oil platforms. Capable of drawing 200,000 barrels a day from the seafloor, Atlantis is located only 150 miles off the coast of Louisiana, in waters nearly 2,000 feet deeper than BP drilled at Deepwater Horizon. According to Congressional documents, the platform lacks required engineering certification for as much as 90 percent of its subsea components—a flaw that internal BP documents reveal could lead to "catastrophic" errors.

President Obama claimed in a Rose Garden ceremony that he was "closing the loophole that has allowed some oil companies to bypass some critical environmental reviews."

But the Department of Interior's Minerals Management Service signed off on at least five new offshore drilling projects soon after June 2011, when the agency's acting director announced tougher safety regulations for drilling in the Gulf. Defenders of Wildlife and the Southern Environmental Law Center filed suit in federal court challenging the MMS's approval of 198 new deepwater leases in the central Gulf after the BP spill began.

The lease sales—an earlier step, before oil companies submit drilling plans—created an incentive to continue offshore drilling despite the risks, attorneys argued. If federal regulators opt to cancel a lease once it's issued, the government must repay the company the fair market value of the lease or compensate it for the cost of its bid plus interest, the lawyers said.

President Obama boasted that he had secured a promise from BP to set aside $20 billion in escrow to compensate those affected by the oil. But that represents a fraction of the damage done.

Furthermore, *The Washington Post* reported that before Obama met with BP officials, "Behind the scenes, the company had signaled what it expected from [the] meeting—and the company appears to have gotten exactly what it wanted." The "escrow account" in 2010 was not $20 billion. BP was to have put in $3 billion in the third quarter of 2010 (ending September 30) and another $2 billion in the fourth quarter (ending December 31, 2010). Thereafter, it was to have made installments of $1.25 billion each quarter for the subsequent three years.

This meant that the necessary money was not available to pay the tens of billions in losses that were real and immediate. It also means that people and businesses had to get in line. This did enormous damage to families and businesses along the Gulf Coast.

The real number for the escrow account in 2010 was $5 billion—months after the oil spill. To put this in perspective, BP had been bringing in between $26 billion and $36 billion annually in profits on revenue of $250 billion, and was paying out more than $10 billion in dividends yearly.

According to a report in *Forbes* at the time, BP could absorb $35 billion in spill costs before it would have a "material impact" on its operations. But instead, thanks to Obama, it was allowed a paltry $5 billion a year, in an installment plan over four years.

However, even this lenient financial arrangement did not go as planned. More than five years were to pass—until October 2015—before BP agreed to a $20 billion settlement with the U.S. Justice Department. Under the agreement, affected parties would receive payment from British Petroleum *by 2019*. So at least nine years of delay. Some wondered if BP was merely stalling so victims died in the meantime. Many did.

Meanwhile, U.S. District Judge Carl Barbier ruled BP was guilty of gross negligence and willful misconduct. He described BP's actions as "reckless." He said the actions of BP's partners, Transocean and Halliburton, were "negligent." Although five people were charged with federal crimes, none of the charges against individuals resulted in any prison time, and no charges were levied against upper level executives. Surprised?

As the Downs Law Group of Coral Springs, Florida, explains it,"Millions of people depend on the Gulf of Mexico for transportation, recreational businesses, commercial fishing and other uses. The inability to use the area surrounding the oil spill caused serious economic distress for many industries, and the impact on the commercial fishing community has yet to be fully comprehended. Those who lived near the shore or on the bayous surrounding the Gulf of Mexico have endured costly property damage and are at increased risk for a variety of medical conditions." The law firm says "countless people are still dealing with the effects of the 2010 disaster" of the BP oil spill.

BP should have been charged with cruelty to animals

The criminal investigation of British Petroleum and those who signed off on the drill-site inspection sheets and safety assurances showed willful fraud and deception. But there is one additional set of criminal charges that should have been added to the list: cruelty to animals. For this was the largest case of cruelty to animals in U.S. history.

"Oil spills are extremely harmful to marine life when they occur and often for years or even decades later," explains Jacqueline Savitz, a marine scientist and climate campaign director at Oceana, an environmental group. The spill area consists of 8,332 species.

Thousands of migratory birds, during a two-to-three-week rest in the Gulf Coast barrier islands on their flight north from South America, died. Birds such as brown pelicans—recently taken off the endangered species list—the Piping Plover and seagulls continued to dive through the oil-soaked ocean to get to the food supply. They were also exposed to it as they floated on the water's surface. Oiled birds can lose the ability to fly, dive for food or float on the water which could lead to drowning. Oil interferes with the water repellency of feathers and can cause hypothermia in the right conditions.

Many volunteers assisted conservation groups to try to clean as many oil-soaked birds and animals as possible. Here's a before and after testimony to their efforts.

As birds groom themselves, they can ingest and inhale the oil on their bodies. While ingestion can kill animals immediately, more often it results in lung, liver and kidney damage which can lead to a slower death.

Manatees and dolphins needed to come up through the oil slick for air; eye irritations are the least of the problems they encountered. For about 5,000 dolphins, the oil spill occurred during their birthing season; mothers who survived had oil on their teats; many of their calves died from lack of nutrition or from ingesting the oil.

Savitz says "Turtles have to come to the surface to breathe and can be coated with oil or may swallow it." Sea turtles such as loggerheads and leatherbacks were impacted as they swam to shore for nesting activities."

Scavengers such as bald eagles, gulls, raccoons and skunks were also exposed to oil by feeding on carcasses of contaminated fish and wildlife.

Oil can be toxic to shellfish including bottom dwelling (lobsters, crabs, etc.) and intertidal (clams, oysters, etc.) species. The Gulf Coast is home to about two-thirds of American oysters.

Fish can be impacted directly through uptake by the gills, ingestion of oil or oiled prey, effects on eggs and larval survival, or changes in the ecosystem that support the fish. Adult fish may experience reduced growth, enlarged livers, changes in heart and respiration rates, fin erosion, and reproduction impairment when exposed to oil. It has the potential to impact spawning success as eggs and larvae of many fish species are highly sensitive to oil toxins. Even the fish which may have survived by staying below the spill, were affected by the oil.

The damaged areas of the Gulf are also the spawning grounds for tuna, marlin, and swordfish. The Gulf is one of only two nurseries for bluefin tuna, more than 90 percent of which return to their place of birth to spawn.

Marine algae and seaweed respond variably to oil, and oil spills may result in die-offs for some specifies.

The BP oil spill still poses risks to YOU

As dramatic as this story is, do you think it doesn't affect you? Think again. Even its aftereffects could kill you. Toxic rain, long-term contamination of our food supply, air pollution, and serious illnesses all make this the biggest threat to the U.S. since the Japanese bombed Pearl Harbor.

The oil dispersant Corexit, previously only used as a surface application, was released underwater in unprecedented amounts. The oil

Dolphins leaping up through the oil-covered surface of the sea.

and dispersant mixture permeated the food chain through zooplankton—from which it proceeded to spread across the entire ecosystem. Chemicals from the spill were found in migratory birds as far away as Minnesota, with a devastating effect on marine wildlife. A 2016 study reported that 88 percent of baby or stillborn dolphins within the spill area "had abnormal or underdeveloped lungs," compared to 15 percent in other areas.

The basic building blocks of life in the ocean have been altered, indicating that the ocean still hasn't recovered from the oil spill.

More than 2,000 historic shipwrecks spanning 500 years of history, rest on the Gulf of Mexico sea floor. Shipwrecks serve as artificial reefs and hotspots of biodiversity by providing hard substrate, something rare in deep ocean regions. Because crude oil was deposited on the sea floor, it affected microbes on the shipwrecks. A recent study shows microbes are still struggling to recover, and since they are affected, the entire food chain that's built upon them is also affected. There's a good chance we have still yet to see all the far-reaching consequences of this event, says University of Southern Mississippi microbial ecologist Leila Hamdan.

And now comes word Donald Trump is lifting the safety precautions Barack Obama imposed after the BP oil spill. The changes are estimated to save the oil industry $824 million over the next decade. Money is more important to the current administration than public safety, obviously. Bob Deans of the Natural Resources Defense Council says "it will put our workers, waters and wildlife at needless risk. That's irresponsible, reckless and wrong."

10.
Land Degradation, Biodiversity Loss and Climate Change

ONLY THE MOST MYOPIC AMONG US CAN FAIL TO SEE THE EVIdence that our climate is changing in ways that imperil our very existence. Some of these changes are occurring gradually, and may not come to full force until the current generation passes. But other effects are here, right now, and are impacting cities and rural areas across America and the world.

The Fourth National Climate Assessment, the most comprehensive report on climate change science ever released—running 1,600 pages—confirms that climate change is already harming Americans and if nothing is done, will devastate the economy and negatively touch millions of lives. The comprehensive report details the many ways in which global climate change is already affecting American communities, from hurricanes to wildfires to floods to drought.

Compelling new evidence shows we will speed past a dangerous climate-risk threshold as soon as 2030 if greenhouse gas emissions continue at their current rate, potentially triggering climate change on a scale that would present grave dangers to much of the living planet.

President Donald Trump couldn't stop the peer-reviewed assessment by the U.S. government's climate scientists, so the White House released it on the afternoon of Black Friday—the busiest post-Thanksgiving shopping day of the 2018 Christmas season. Then he dismissed the findings of the report, remarking "I don't believe it."

"We don't care," said Steven Milloy, a climate denier who was on Trump's Environmental Protection Agency transition team. Milloy, who also rejects the science behind second-hand smoke is, unsurprisingly, a paid advocate for the tobacco industry.

Trump disparages climate science because it interferes with his desire to curry favor with the powerful fossil fuel industry, in addition to apparently wanting to follow Vladimir Putin's objectives.

Climate change has caused extreme and more frequent events.

With Trump, it's always about money and greed.

Besides the federal government's assessment of the global warming threat, a new three-year, United Nations-backed landmark study released in May 2019 says that nature is in freefall and the planet's support systems are stretched so thin that we face species extinctions and mass human migrations.

The 8,000-page report was compiled by more than 500 experts in 50 countries, and is the greatest attempt yet to assess the state of life on earth. It says that tens of thousands of species are at high risk of extinction, countries are using their natural resources at a rate that far exceeds their ability to be renewed, and nature's capacity to contribute food and fresh water to a growing human population is being compromised in every nation on earth.

The latest UN report lists the cutting down of forests, the overfishing of seas, the reckless exploitation of land, and the pollution of air and water as factors contributing to the rapid decline of our natural world. Industrial farming is also a culprit.

On the American continent alone, nature is said to underpin economies with "free services" like free clean water, air and the pollination of all major human food crops by bees and insects, and contribute more than $24 trillion a year.

Sir Robert Watson, chair of the study—who also led programs at NASA—told *HuffPost* that "we are at a crossroads. The historic and current degradation and destruction of nature undermine human well-being for current and countless future generations... Land degradation, biodiversity loss and climate change are three different faces of the same central challenge: the increasingly dangerous impact of our choices on the health of our natural environment."

An obsession with economic growth as well as a swelling human population are also factors driving the destruction. Gross Domestic Product (GDP) is expected to nearly double by 2050 and the population in the Americas is expected to increase 20 percent to 1.2 billion over the same period.

Climate-change deniers are ignoramuses or deliberately obtuse

Anybody who tries to debunk the reality of global warming and its implications for the present and future generations is on a par with those in the past who insisted the earth was flat. Or they are embracing denial for political reasons. Even the U.S. military officially views climate change as a threat multiplier, one that is likely to worsen already existing weaknesses of government and poverty.

Forbes magazine writes: "Thankfully, major U.S. manufacturers are already aware of the problem... Apple detailed how severe weather brought on by climate change can disrupt the production or availability of components. General Motors talked about a 'zero-emissions future' in its 2017 sustainability report. And Cardinal Health has said it is 'actively involved' with the EPA's SmartWay program to improve fuel efficiency and cut back on greenhouse gas emissions."

Donald Trump has led the pack in disparaging the science behind climate change, calling it "an expensive hoax," trying to obfuscate the issue and convince people that there is widespread dispute over the reality of the problem. But as with so many things, he's lying.

White House spokeswoman Sarah Huckabee Sanders said "even Obama's Undersecretary for Science didn't believe the radical conclusions of the report that was released." This appears to be a reference to Steven Koonin, a former chief scientist for BP, who spent 18 months in Obama's energy department. Koonin has long maintained climate science is not "settled," but of course he worked for a company that inflicted the biggest manmade disaster on the U.S. in history.

Also in Trump's corner are entities with a suspicious agenda. For example, the lobbying group, the Alliance of Automobile Manufacturers, also cast doubt on the negative effects of tailpipe pollution on human health. Anybody who holds a similar view should look at old pictures of Los Angeles smog as evidence that car and truck exhaust is an obvious health hazard.

That humans are causing global warming is the position of the

Academies of Science from 80 countries plus many scientific organizations that study climate science. More specifically, around 95 percent of active climate researchers actively publishing climate papers endorse the consensus position.

A new poll by Yale and George Mason University released in January 2019 shows that most Americans have rejected Trump's position and that they believe climate change is a real and present danger.

Subtle changes we may not notice

Climate change can affect human health in two main ways: first, by changing the severity or frequency of health problems that are already affected by climate or weather factors; and second, by creating unprecedented or unanticipated health problems or health threats in places where they have not previously occurred.

Days that are hotter than the average seasonal temperature in the summer or colder than the average seasonal temperature in the winter cause increased levels of illness and death by compromising the body's ability to regulate its temperature or by inducing direct or indirect health complications.

Changes in the climate affect the air we breathe, both indoors and outdoors. The changing climate has modified weather patterns, which in turn have influenced the levels and location of outdoor air pollutants. Finally, these changes to outdoor air quality and aeroallergens also affect indoor air quality as both pollutants and aeroallergens infiltrate homes, schools, and other buildings. Poor air quality, whether outdoors or indoors, can negatively affect the human respiratory and cardiovascular systems. Higher pollen concentrations and longer pollen seasons can increase allergic sensitization and asthma episodes and thereby limit productivity at work and school.

Higher concentrations of CO_2 stimulate growth and carbohydrate production in some plants, but can lower the levels of protein and essential minerals in a number of widely consumed crops, including wheat, rice, and potatoes, with potentially negative implications for human nutrition.

Mental health consequences of climate change range from minimal stress and distress symptoms to clinical disorders, such as anxiety, depression, post-traumatic stress, and suicide.

Warnings from the EPA

Our government's own Environmental Protection Agency says that "increases in the frequency or severity of some extreme weather events, such as extreme precipitation, flooding, droughts, and storms, threaten the health of people during and after the event. The people most at risk include young children, older adults, people with disabilities or medical conditions, and the poor. Extreme events can affect human health in a number of ways." The EPA list is as follows:

• Reducing the availability of safe food and drinking water.

• Damaging roads and bridges, disrupting access to hospitals and pharmacies.

• Interrupting communication, utility, and health care services.

• Contributing to carbon monoxide poisoning from improper use of portable electric generators during and after storms.

• Increasing stomach and intestinal illness, particularly following power outages.

• Creating or worsening mental health impacts such as depression and post-traumatic stress disorder (PTSD).

Many Americans have already experienced these ill effects.

A fuller explanation of climate change

Sea levels are rising and oceans are becoming warmer. Longer, more intense droughts threaten crops, wildlife and freshwater supplies. From polar bears in the Arctic to marine turtles off the coast of Africa, our planet's diversity of life is at risk from the changing climate.

Climate change poses a fundamental threat to the places, species and people's livelihoods Americans (and many others) wish to protect. To adequately address this crisis we must urgently reduce carbon pollution and prepare for the consequences of global warming, which we are already experiencing.

Greenhouses gases, such as carbon dioxide, trap heat in the atmosphere and regulate our climate. These gases exist naturally, but humans add more carbon dioxide by burning fossil fuels for energy (coal, oil, and natural gas) and by clearing forests. Greenhouse gases act like a blanket. The thicker the blanket, the warmer our planet becomes. At the same time, the earth's oceans are also absorbing some of this extra carbon dioxide, making them more acidic and less hospitable for sea life.

The increase in global temperature is significantly altering our planet's climate, resulting in more extreme and unpredictable weather. As climate change worsens, dangerous weather events are becoming more frequent or severe. People in cities and towns around the United States are facing the consequences, from more frequent heat waves and record droughts and wildfires to coastal storms and flooding.

Scientists in the United States and the world have reached an overwhelming consensus that climate change is real and caused primarily by human activity. Respected scientific organizations such as the National Academy of Science (NAS), the Intergovernmental Panel on Climate Change (IPCC) and World Meteorological Association (WMO) have all identified climate change as an urgent threat caused by humans that must be addressed.

According to the National Wildlife Foundation (NWF), "Higher temperatures will lead to drier conditions in the Midwest's Prairie Pothole region, one of the most important breeding areas for North American waterfowl. Sea-level rise will inundate beaches and marshes and cause erosion on both coasts, diminishing habitat for birds, invertebrates, fish, and other coastal wildlife. Higher average temperatures and changes in rain and snow patterns will enable some invasive plant species to move into new areas. Insect pest infestations will be more severe as pests such as mountain pine beetle are able to take advantage of drought-weakened plants. Pathogens and their hosts that thrive in higher temperatures will spread to new areas."

The National Geographic Society has outlined the effects of climate change as follows:

• Ice is melting worldwide, especially at the earth's poles. This includes mountain glaciers, ice sheets covering West Antarctica and Greenland, and Arctic sea ice. In Montana's Glacier National Park the number of glaciers has declined to fewer than 30 from more than 150 in 1910.

• Much of this melting ice contributes to sea-level rise. Global sea levels are rising, and the rise is occurring at a faster rate in recent years.

• Rising temperatures are affecting wildlife and their habitats. Vanishing ice has challenged species such as the Adélie penguin in Antarctica, where some populations on the western peninsula have collapsed by 90 percent or more.

• As temperatures change, many species are on the move. Some butterflies, foxes, and alpine plants have migrated farther north or to higher, cooler areas.

• Precipitation (rain and snowfall) has increased across the globe, on average. Yet some regions are experiencing more severe drought, increasing the risk of wildfires, lost crops, and drinking water shortages.

• Some species—including mosquitoes, ticks, jellyfish, and crop pests—are thriving. Booming populations of bark beetles that feed on spruce and pine trees, for example, have devastated millions of forested acres in the U.S.

• Sea levels are expected to rise between ten and 32 inches or higher by the end of the century.

• Hurricanes and other storms are likely to become stronger. Floods and droughts will become more common. Large parts of the U.S., for example, face a higher risk of decades-long "megadroughts."

• Less freshwater will be available, since glaciers store about three-quarters of the world's freshwater.

• Some diseases will spread, such as mosquito-borne malaria (and the 2016 resurgence of the Zika virus).

• Ecosystems will continue to change: Some species will move farther north or become more successful; others, such as polar bears, won't be able to adapt and could become extinct.

Sea level rises pose an urgent danger

David Introcaso of *Healthaffairs.org* writes that "The current rate of rise in Global Mean Sea Level is greater than any time in at least 2,800 years. As for rising sea levels from ice melt, should Greenland ice sheets thaw in their entirety they would add 20 feet to the height of global seas. The thaw of the West Antarctic Ice Sheet, that is presently vanishing faster than any previously recorded time, would add another ten feet. At 20 feet, most of Florida and a third of New York City would be under water. An estimated 145 million people worldwide live three feet or less above sea level and ten percent of the world's population or nearly 800 million live less than 30 feet from present sea levels. Eleven of the 16 megacities, those with more than 15 million people, are built on coasts, for example, Jakarta, Los Angeles, Manila, Mumbai, Osaka, Shanghai and Tokyo.

"Rising seas or flooding compromises drinking water, human waste water treatment and storm water disposal that, in turn, results in increased risk of waterborne diseases caused by pathogens such as bacteria, viruses and protozoa. Between 1948 and 1994, 68 percent of waterborne disease outbreaks in the U.S. were preceded by extreme precipitation events," David Introcaso reports.

Florida's cities, infrastructure, beachfront homes, and natural ecosystems are among the most vulnerable in the world to sea-level rise. A headline from the *Miami Herald* says it all: "FLORIDA HAS MORE TO LOSE WITH SEA RISE THAN ANYWHERE ELSE IN THE U.S., NEW STUDY SAYS." In the Sunshine State, almost 1.8 million people live near the water.

Former Florida State Senator Ellyn Bogdanoff tells a fascinating story about South Florida and its vulnerability. She explains that the area "has an incredibly complex flooding control system with more than a thousand miles of levees, over 700 miles of canals, and nearly 200 water control structures designed to keep us dry and protect our drinking water. We take this system for granted, but rising sea levels are challenging the usefulness of the system.

"To understand how we got here, we have to look back nearly a century. The devastation of the Hurricanes of 1926 and 1928 pushed the federal government to act and the Herbert Hoover Dike around Lake Okeechobee was authorized in 1930 as a result.

"It took just seven years to complete the dike, and it was designed to ensure the flooding during the hurricane of 1928 (that claimed nearly 2,500 lives) never happened again."

Despite the flood control efforts, she emphasizes, "Persistent rain caused significant flooding in 1947 and almost 90 percent of southeastern Florida found itself underwater. Stories during the flooding made the region seem like the Wild West, with several feet of water taking months to dissipate in some areas."

Southeast Florida fixed its flooding problems, for the time being, and has not seen flooding of that magnitude since. But Hurricane Andrew in 1992 nevertheless dealt the area a devastating blow.

Now, Sen. Bogdanoff says, "Sea level rise has brought a whole new challenge to our region and to our state. We are watching the slowest disaster movie ever recorded and we are barely past the opening credits. However, this does not diminish our need to plan and act now."

She concludes, "Due to our limestone bedrock, a giant wall like the Great Wall of China will not keep the waters out. The water is coming from all directions, including below. We need innovative and unique solutions that will take some time to develop and deploy. But I am hopeful."

A Union of Concerned Scientists (UCS) report took a flood model from the National Oceanic and Atmospheric Administration and applied it to housing information from the Zillow real estate database. The findings were alarming.

"By 2045, nearly 64,000 homes in Florida face flooding every other week. Half of those are in South Florida," the *Miami Herald* writes. By the end of the century, Florida's number of at-risk homes jump from 64,000 to a million. "That puts the Sunshine State at the top of the list nationwide for homes at risk," the newspaper concludes.

Founded in 1565 by Spanish explorers, St. Augustine, Florida is the oldest continuously inhabited European-established settlement within the borders of the continental United States. From living history museums like the Castillo de San Marcos to Gilded-Age hotels, swash-buckling adventures and specialized tours of the historic district that draw six million visitors a year from all over the world, this quaint city has tons of appeal. But St. Augustine is increasingly vulnerable to flooding and sea level rise.

One family has moved seven times in two years. Hurricanes Matthew and Irma—only five months apart—flooded homes with salt water, half-treated sewage, and debris carried by raging water. But even rain storms pose problems.

The Union of Concerned Scientists says "Many people in St. Augustine don't realize how vulnerable their city is. The increasing pace of sea level rise could overwhelm their ability to stay. The low-lying city is prone to flooding, storm surges, and erosion. And its aging infrastructure takes a beating even in light rainstorms, with nuisance flooding regularly closing streets in the historic districts. For example, a rise of three feet [in sea level]—which is well within current projections—could permanently inundate portions of the city's historic districts. Meanwhile, storm surges and flooding could undermine the foundations of many historic buildings unless protections are put in place."

A lot of the oldest buildings in St. Augustine are constructed with coquina, an early form of concrete made from sea shells. A 19th-

The nation's oldest city, St. Augustine, Florida is vulnerable to sea level rises.

century coquina seawall protects the downtown from flooding, but the ocean waves' pounding of the wall during tropical storms has caused its deterioration and created the need for reinforcement. The UCS says "In 2011, the National Park Service built a 'living' seawall. Over time, sediment will build up in front of the wall, and marine life and vegetation will establish themselves there, protecting the historic seawall from erosion and creating a natural habitat." But that's still years away.

With Republicans in the Florida legislature closing their ears and their minds anytime anyone mentions climate change, funds are hard to come by. But the city's Mayor, Nancy Shaver, traveled to the Netherlands to speak on a panel about vulnerability to rising sea-levels. On her last day there, she visited Oosterscheldekering, the largest of the 13 dams and storm surge barriers, designed to protect the Netherlands from flooding from the North Sea. There she stood with Gert-Jan Schotmeijer, a senior advisor at Deltares—one of the world's leading research institutions on water management. When she showed him St. Augustine's sea-level rise map for 2050, Schotmeijer said, "That looks like Jakarta." He offered to study St. Augustine's data, to validate their flood maps and provide a road map for prevention—and consult with the city at a price within their limited budget.

Says the Climate Reality Project, "The findings offer a stark reminder that without prompt, serious action to transition our global economy away from the fossil fuels driving the climate crisis toward clean, renewable energy like wind, solar, and geothermal, hundreds of thousands of people in Florida (and millions more around the world) may find their homes swamped by mid-century."

Other areas of the U.S. are also at risk to sea level rise

With some of the most populated and desirable locations in America located near the sea, just how many people will be displaced when the sea levels rise?

Forbes magazine analyzed scientific data and concluded that besides Florida, the U.S. cities most at risk to global flooding are New York, Seattle and San Francisco, the nerve centers of the financial and technology industries. San Diego also makes the top ten list of flood-risk cities.

New York City, a place completely surrounded by water has 346,000 people living near the coast. In 2012, the city's subways flooded during a major storm, Hurricane Sandy. Some of the most notable cities in America will be hugely impacted by global warming if projections of higher sea levels hold true.

Surprisingly, according to *Forbes*, "the two areas with the most expensive properties that are at risk to rising sea levels are both in Connecticut: Riverside and Darien are both high risk zones where the median home costs almost $2 million."

Twenty-three of the 50 states in America have some level of their population living in coastal zones. California and Washington, both of which abut the Pacific ocean rank as the places with the second and third most people at risk to rising sea levels.

The New Yorker magazine reports that Louisiana loses a football field's worth of land every hour and a half, and that engineers are in a race to prevent it from sinking into oblivion. Elizabeth Kolbert writes that "Since the days of Huey Long, Louisiana has shrunk by more than 2,000 square miles. If Delaware or Rhode Island had lost that much territory, the U.S. would have only 49 states." This is attributable not just to hurricanes, but also attempts by the Army Corps of Engineers to reroute the Mississippi river and higher-than-normal rainfall.

Weather has gotten more extreme

Many states up and down the Atlantic and Gulf coasts, have witnessed an increase in extreme weather events as a result of our changing climate. The reason why is fairly easy to comprehend.

"For a long time, we've understood, based on pretty simple physics, that as you warm the ocean's surface, you're going to get more intense hurricanes," world-renowned climate scientist Dr. Michael Mann told Climate Reality. "Whether you get more hurricanes or fewer hurricanes, the strongest storms will tend to become stronger."

So as sea surface temperatures become warmer, hurricanes become more powerful. In recent years, with temperatures in parts of the Atlantic Ocean running well above longtime averages, we've seen how this dynamic can play out across the southeastern United States. And Florida's geography—that of a relatively flat peninsula, jutting out into the Atlantic Ocean and Gulf of Mexico—leaves it especially vulnerable.

While Hurricane Matthew in October 2016 ultimately stayed just offshore, traveling parallel to the Florida coastline, the telltale signs of the climate crisis were all over the speed with which it strengthened, leading to devastation on the Caribbean island of Haiti.

"It spun up from a tropical storm into a Category 5 hurricane in just 36 hours. That's extremely unusual," Climate Reality's founder and chairman, former U.S. Vice President Al Gore, said of Matthew.

In September 2017, Hurricane Irma became the first Category 4 hurricane to strike Florida since Charley in 2004, prompting the largest evacuation in the state's history—one million residents and tourists.

The New York Times reported that "Hurricane Michael [in October 2018] took millions of residents by surprise, intensifying from a tropical storm to a major hurricane in just two days and leaving little time for preparations. Part of the explanation for the intensification was warmer-than-average waters in the Gulf of Mexico."

As it approached the Florida Panhandle, Michael attained peak winds of 155 m.p.h. just before making landfall. Then it veered north-easterly across Georgia and up toward Chesapeake Bay, before exiting off the Mid-Atlantic states.

Climate change is also spawning more tornadoes and in new areas. Although persons living in "tornado alley"—an area between the Mississippi and Missouri rivers in the Midwest—are all too accustomed to the sound of sirens going off and dark, swirling clouds descending to do major damage, some of the deadliest tornadoes of late have occurred elsewhere.

One such twister hit Alabama and Georgia in March 2019, with winds of 170 m.p.h. These storms fit into a growing trend for a region meteorologists are calling "Dixie Alley" in the Southeastern United States to have more tornadic activity than it has decades before. Unlike hurricanes, weather forecasters cannot typically predict where a tornado will strike until 13 minutes before it hits.

❑

U.S. efforts to avoid what the world's leading climate scientists increasingly describe as total dystopia remain anemic. Besides Donald Trump's efforts to debunk climate change, recent state efforts to limit greenhouse pollution via increased dependence on renewables, a ban on new drilling and a carbon tax failed respectively, in Arizona, a California county and in Washington largely due to the fossil fuel industry spending over $60 million in opposition.

Yet McClatchy newspapers reported in April 2019 that the Trump re-election campaign is worried that the Democrats' success in the mid-term elections shows a growing voter concern over global warming. So Trump will point to U.S. greenhouse gas emissions that decreased in his first year in office and attribute it to private sector innovation—not policies put in place by President Obama. The assertion will be that Trump has been more effective in cutting emissions from major industrial sources, even though that's laughably absurd.

11.
The Folly of Trump's Border Wall
...and A Better Solution

THE BIBLICAL ACCOUNT TELLS US THAT JESUS WAS BORN IN A manger in Bethlehem because there was no room at the inn. After the birth of the baby Jesus, an angel of the Lord appeared to Joseph in a dream and told him to flee to Egypt because Jesus's life was in danger from King Herod —a man who had inherited wealth, and in whom there was a dark and cruel streak in his character, that was fed by intrigue and deception that afflicted everyone around him.

Upon the Holy Family's arrival at Egypt, there is no record of them having been turned away by heavily armed border patrol agents, or having their child forcibly separated from them just because they were seeking a safe haven and were of a different nationality. So baby Jesus did find shelter as a refugee in Egypt. After King Herod died, the Holy Family left Egypt and went first to Judaea, although upon discovering the new king was also violent and aggressive, they then fled to Galilee, which had become a center for refugees.

Counting the trips made during Passover, Mary—the mother of Jesus—walked an estimated 4,000 miles on foot over the years; there is no record of her ever having ridden a donkey, even while pregnant!

It is hard to imagine that these long journeys on foot continue to be made by people in 2019, but in fact there are upwards of 60 million worldwide who have been forced by violence and conflict to flee their homes and walk hundreds if not thousands of miles to seek safety in other countries. That's more than live in the U.K. or France. If all these migrants were a country, they'd be the 21st most populous nation in the world. More than half are children. This crisis is the worst it has been since World War II.

In Syria, millions of people have been forced from their homes by the terrorist group ISIS. In Afghanistan, ongoing violence has created one of the largest, longest-lasting refugee crises. Millions of South Sudanese have fled to Uganda and other surrounding nations as a civil war rages. In Myanmar, the minority has faced deadly persecution and many have fled. In Somalia, a combination of natural disasters and 25 years of conflict have produced hundreds of thousands of Somali refugees.

An anti-immigrant strain of populism has afflicted the United States and other countries, making it more difficult for international organizations and humanitarian groups to resettle and care for refugees, especially in Europe and the U.S. The problem is so huge, it won't go away simply by ignoring it or trying to inflict as much cruelty as possible on these fleeing minions as they approach our borders. Millions are suffering from poverty and persecution and natural disasters and given the wealth of this world, it is a moral responsibility—as the Good Book admonishes—not to mistreat them: "The foreigner residing among you must be treated as your native born. Love them as yourself, for you were foreigners in Egypt."

The Trump administration is processing only 60 applicants for asylum per day, and has closed points of entry, and this has pushed impoverished migrants into desperation and dangerous situations. A little girl died of thirst as border patrol agents were seen taking jugs of water that had been left by Good Samaritans to help thirsty migrants and emptying those jugs on the ground to deprive people of life saving hydration. All of this at the direction of our government. Incidentally, people have a right to seek asylum under U.S. law even if they cross the border illegally.

The problem is that there is a backlog of pending asylum claims. The number of cases have increased by nearly 50 percent since Mr. Trump took office in 2017, and former Attorney General Jeff Sessions deliberately exacerbated the problem by moving to reopen more than 300,000 cases that had previously been closed. This has stretched the court docket to more than one million pending cases.

Trump has deliberately worsened the border problem

During his campaign for President, Donald Trump promoted nativist racism with his constant scapegoating of migrants fleeing violence and poverty in their home countries and approaching our southern border with Mexico.

A more legitimate issue would have been if he had focused on the fact that since the 1990s, drug cartels have literally taken over some

A caravan traveling through Mexico on foot headed for the U.S. border.

American cities in states bordering Mexico, infiltrating local governments, corrupting police departments, and exposing U.S. citizens to acts of violence. That problem has been under-reported by the U.S. media for years, and neglected by a succession of Presidents, despite the fact that this constitutes a true invasion of our country. It's far easier to pick on homeless refugees, and portray them as a threat to our national security than it is to declare war on actual drug kingpins.

Once he took office, Trump immediately declared a "zero tolerance" policy toward immigrants and claimed "we're talking about an invasion of our country with drugs, with human traffickers, with all types of criminals and gangs."

The Trump administration's efforts to claim that thousands of terrorists or potential terrorists have tried to cross into the U.S. have been roundly mocked and debunked. Only six known or suspected terrorists were caught trying to come to the U.S. in the first six months of 2018.

Similarly, Trump likes to characterize the border as a haven for drug smuggling. It's true that a lot of drugs consumed in the U.S. come in via the border with Mexico. But most drugs enter through Border Patrol checkpoints, and also by planes landing in various American cities—not through the open land where walls might be built.

Trump's lies versus the truth about the migrants

Prior to the 2018 mid-term elections Trump made it sound like America was going to be overrun by hundreds of thousands of criminals who would maraud through the streets, committing acts of violence, and tearing our country apart.

Rightwing media claimed these individuals were even traveling in luxury motor coaches or other conveyances, being regularly fed along the way, and were probably financed by the conservatives' favorite whipping boy, liberal billionaire philanthropist George Soros. (Mr. Soros, incidentally, has been idiotically vilified as an antisemite even though he is a Jew who survived Nazi-occupied Hungary).

The truth is, the vast majority of migrants were families who in many cases had walked 2,500 miles—in flip-flops or flimsy footwear, some of them barefoot—from Central America. They were baked red from walking in the sun and heat, feet bloodied with blisters, stripping down to bathe in murky waters enroute. They crossed rivers swollen by increased flows caused by intense rains and faced off against riot police as they made their way northward.

Many did travel in "caravans," purely as a matter of safety. Travel by caravans goes back as far as recorded history (see, for example, Genesis 37:25), though today's migrants didn't have the donkeys, camels or ox-carts used in Biblical times. The lucky ones might be able to hitch a ride for a few miles on a flatbed truck, but trains have stopped running in much of Mexico so they couldn't ride the rails.

In any case, there is a legal obligation to hear asylum claims

A small child surrounded by gun wielding agents is part of Trump's legacy.

from migrants who have arrived in the U.S. if they say they fear violence in their home countries—which is a credible claim on their part, given the deplorable conditions in their home countries.

Those seeking asylum must be fleeing due to a serious fear of persecution. Under international law, they are considered refugees.

But in a sadistic plan apparently hatched in the White House by aides like Stephen Miller, and enthusiastically carried out by Attorney General Jeff Sessions, the federal government began forcibly separating children as young as babies from their parents, without setting up any tracking method for eventually reuniting them. Thousands of children were literally ripped from their parents' arms, such as the Honduran woman who was breastfeeding her infant when authorities took her baby away. Parents were often tricked into giving up their children, and quickly deported while thousands of kids stayed behind, or got lost in a jumbled system of temporary foster care.

The spectacle of barefoot children, some in diapers, choking on tear gas lobbed at them by U.S. border patrol agents, or sleeping on the rough asphalt as they waited interminably under the scorching sun, shocked the conscience of the nation. So, too, did the incarceration of these innocents in wire cages, set up in warehouses, an old Wal-Mart and even under a bridge in Texas. They were given only foil blankets to sleep on. Heavily armed guards and other staff were instructed to not even comfort a crying child, no matter how young and distraught.

Then there was the bizarre scenario of pre-school aged children, who do not even speak English, being hauled before judges who were deciding their fate.

Jennifer Podkul, director of policy at Kids in Need of Defense, which represents children in immigration court, told the *Tampa Bay Times* "They're willing to risk harm to a child being traumatized, separated from a parent and sitting in federal detention by themselves, in order to reach a larger policy goal of deterrence."

Such cruelty provoked condemnation from Republicans as well as Democrats, but First Lady Melania Trump compounded the controversy when she inexplicably wore a jacket emblazoned with the slogan "I don't really care, do U?" on the back, as she went to Texas to visit a detention facility where children were being held.

Ironically, Melania Trump's own parents were recently sworn in as U.S. citizens—and they relied on the same immigration process Donald Trump wants to end. The immigration lawyer who helped the First Lady, Michael Wildes, told CNN that the President's attacks on immigration were "unconscionable."

The President's daughter, Ivanka Trump, said in one interview, "I am a daughter of an immigrant. My mother [Ivana] grew up in communist Czech Republic. She came to this country legally."

But that's not the whole story—according to the former Mrs. Trump herself. In her autobiography, *Raising Trump*, Ivana Trump admits she entered into a fraudulent marriage in 1971 solely to acquire an Austrian passport that would allow her to immigrate to Canada.

In the U.S., marrying for immigration purposes would be a felony called "marriage fraud." According to ICE, marriage fraud "carries a prison sentence of up to five years and a fine of up to $250,000, and applies to both foreign nationals and U.S. citizens who perpetrate this crime." Convenience marriages contracted in order to obtain immigration status are illegal also in Austria and Canada.

Once in Canada, Ivana regularly crossed the border into the U.S. to work as a model, but it remains unclear whether she acquired a U.S. working permit to do so. On one such trip, she met Donald Trump, who proposed marriage. Ivana returned to Canada to pack. Returning to the U.S. on a tourist or work visa, with the intent to marry, would be a big no-no in immigration law. We don't know exactly what happened, but we do know Ivana waited 11 years to get her citizenship after her marriage to Donald.

So three of President Trump's own children—Don Jr., Eric and Ivanka—were born to a woman who was not even a U.S. citizen at the time she gave birth!

Border security officials condemn Trump policies

President Trump has upped the ante by repeatedly threatening to close the U.S.-Mexico border if authorities there fail to stop migrants—an imperious move that would disrupt hundreds of thousands of legal freight, vehicle and pedestrian crossings each day.

In an article for *Politico*, Alan Bersin (who served as commissioner of U.S. Customs and Border Protection and a chief officer for the U.S. Dept. of Homeland Security), Nate Bruggeman (who has held senior policy positions at DHS and Customs and Border Protection), and Ben Rohrbaugh (who was with the National Security Council and also served in senior positions at DHS), declare that Donald Trump "is the worst President for border security in the last 30 years.

"Despite the administration's attempts to shift blame for the chaos, make no mistake: It is Donald Trump himself who is responsible. Through misguided policies, political stunts and a failure of leadership, the President has created the conditions that allowed the asylum problem at the border to explode into a crisis. The solution to our current border troubles lies in reforming the U.S. asylum system and immigration courts and helping Central America address its challenges—not in a 'big beautiful' wall or shutting down the border. Yet effective action on these issues has been missing. And the President has now so poisoned the political well with his approach that there is little hope of meaningful Congressional action until after the next election."

The three immigration experts point to the fact that "The Clinton, Bush and Obama administrations had worked hard to tackle the problem of illegal migration through substantial increases in border security staffing, improvements in technology, innovations in strategy and improved security coordination and assistance to Mexico. Coupled with improved economic conditions in Mexico, these administrations were hugely successful in deterring and breaking the cycle of illegal crossing: Unlawful Mexican economic immigration, which had historically been the primary immigration enforcement issue at the border, dropped nearly 90 percent between 2000 and 2016."

They acknowledge that "the nature of undocumented immigration to the U.S. has changed. Today, it is primarily driven not by Mexican economic migrants—and not by a flood of criminals, as Trump claims—but rather by large numbers of families and minors from Central America who are seeking political asylum."

The three immigration specialists conclude "The President's wall is, in other words, unmoored from operational reality. A wall will not make Central America a better place to live. A wall will not stop asylum seekers from coming to the United States and being able to claim asylum. A wall will not address, let alone fix, the issues with America's asylum system and immigration courts. The President's attacks on Mexico and Central America, coupled with the lack of a coherent strategy for the region, have made harder the already difficult work of addressing the underlying drivers of illegal migration from Central America. Instead of working to address these problems, the President has actively made the problem worse by redirecting resources and attention to his irrelevant wall, antagonizing the people he needs to partner with to actually solve immigration problems, exacerbating backlogs and resource shortages by shutting down the government and announcing enforcement measures that cannot be sustained and which result in increasing numbers of migrants..."

A surprising reason why many migrants are coming

Although a lot of migrants are definitely fleeing violent conditions in their home countries, Robert Albro, a researcher at the Center for Latin American and Latino Studies at American University, says "The main reason people are moving is because they don't have anything to eat. This has a strong link to climate change—we are seeing tremendous climate instability that is radically changing food security in the region."

Migrants don't often specifically mention "climate change" as a motivating factor for leaving because the concept is so abstract and long-term, Albro said. But people in the region who depend on small farms are painfully aware of changes to weather patterns that can ruin crops and decimate incomes.

A study of Central American migrants by the World Food Program last year found that nearly half described themselves as food insecure.

Data shows a surge in outward migration from western Honduras, a prime coffee-producing area, said Stephanie Leutert, an expert in Central American migration and security at the University of Texas.

Coffee used to be worth something, but it's been seven years since there was a decent price, due to a blight and deforestation that has affected the coffee crops.

A third of all employment in Central America is linked to agriculture, so any disruption to farming practices can have devastating consequences.

The U.S. created a lot of the problems in Central America

Almost totally absent from this discussion is any awareness or acceptance of U.S. responsibility for the repressive and corrupt governments that have been the driving force behind this flood of refugees.

Ed Corcoran, a former strategic analyst at the U.S. Army War College, where he chaired studies for the Office of the Deputy Chief of Operations, provides a brief history of how the United States precipitated a lot of the problems in Central America.

"A hundred years ago, American businessmen basically took control of Central America," he declares. "With the mostly white, Spanish-speaking aristocracy in the region, they set up subservient governments that strongly supported U.S. commercial interests at the expense of the indigenous populations. The U.S. government turned a blind eye to, or abetted, this repressive commercial domination of 'banana republics.'

"The situation was exacerbated by the Cold War against Soviet communism. Unfortunately, that struggle was given such overwhelming priority in foreign policy that the United States often supported brutal autocrats so long as they were anti-communist. In Central America, this intensified existing U.S. support for its repressive governments.

"In post-war Guatemala, popular uprisings had brought in reform governments that directly threatened U.S. business interests, leading to a CIA-led invasion that resulted in a bitter civil war. The United States supported a series of repressive governments responsible for widespread massacres in the country, for which President Clinton eventually apologized.

"A UN-sponsored peace accord in 1996 ended the civil war and led to free elections, but resulted in deeply corrupt governments."

Corcoran says that "The situation is no better in El Salvador, which also had bitter civil conflicts in the 1980s. Pre-war political turmoil had the commercial elites strongly supporting the country's military government's brutal suppression of rural resistance... Poor economic conditions and drug trafficking led to the rise of two violent street gangs that the government has been totally unable to suppress.

"The United States established a continuing military presence (in Honduras). As the Honduran army became heavily involved with anti-guerilla activities, a CIA-backed campaign became entangled in a range of extra-judicial killings.

"This involvement deepened in June 2009, when a *coup d'etat* ousted the elected President Manuel Zelaya. The United States declined to insist on Zelaya's return, instead continuing cooperation with the new government. U.S.-trained forces continue to suppress popular demonstrations in the country, and the government continues to favor foreign corporations at the expense of the local population.

"The situation is further complicated by the U.S. demand for illicit drugs—and its prohibition of them—which fuels much of the criminality in the region, as well as helping to destabilize Mexico.

"Overall, the current dismal governance in much of Central America is a direct result of callous U.S. policies," Ed Corcoran bluntly states.

He opines that "President Trump's threat to cut off U.S. aid to the countries refugees flee is exactly the opposite of what's really needed—a comprehensive regional strategy for economic development in support of democracy and human rights."

A sort of mini-Marshall plan, like President Harry Truman implemented to aid European countries in the aftermath of WWII, could help. Upgrading schools and infrastructure would help. Since agriculture remains the driving force of the Central American economies, people there could use assistance with that too, especially in light of the climate change-induced problems.

"The principal emphasis should be on infrastructure, schools and public health plans that support economic expansion with immediate job opportunities, as well as a regional market to expand the Central American economies."

China has also pledged $150 million in support of development in El Salvador, and the World Bank has long been promoting jobs in the region. Former Secretary of State George Shultz has recommended working with the Inter-American Development Bank to redirect its finance focus to its poorest member countries.

The United Nations could also provide support for prosecuting criminals in those countries to reduce the incidence of crime and break up gangs, purge the police force of corrupt officials, and implement new laws for safer streets. Throughout the region, the United States has backed land registration of the disempowered to counter the prevalence of violent, illegal seizures of private property.

As Ed Corcoran insightfully philosophizes, "Instead of spending an estimated $40 billion on an ineffectual border wall (like Trump wants to do), the money would be much better spent promoting prosperity and improving local governance (in Central America) to minimize the number of refugees who are displaced in the first place."

Obviously, there also needs to be significant improvement in the capacity to fairly and swiftly assess the claims of asylum seekers in the United States. Those who meet the criteria should be admitted without delay.

That means more humane conditions for people awaiting a hearing, more interviewers who determine whether migrants meet the "credible fear" threshold, and more immigration judges. The inadequacy of current capacity has contributed to a backlog of some one million cases in the immigration courts.

The border crisis is real, but there are better solutions

The *Chicago Tribune*—a Republican-leaning newspaper—agrees with many of the views analyst Ed Corcoran has propounded. The *Trib* editorialized, "At the end of World War II, much of Europe lay in literal ruins. As many as 20 million people were dead. Millions of survivors were displaced from their home countries. Factories, office buildings, roads and bridges were wrecked. Famine loomed.

"The U.S. government, having played a large role in defeating Germany, could have chosen to declare its job done and leave Europeans to rebuild Europe. Instead, this country opted to restore economic health and prevent political upheaval. Americans had learned the hard way that our security and prosperity were inseparable from events in Europe.

"In 1948, President Harry Truman launched the Marshall Plan, named after Secretary of State George Marshall. Over more than a decade, it provided 16 countries a huge amount of aid—equal to five percent of U.S. gross domestic product, which would be nearly $1 trillion today. The effort did much to bring prosperity and democracy to a bloodied continent. It also helped stem the tide of communism in early years of the Cold War."

The *Tribune* points out that "For years, Central America has endured a humanitarian crisis. It was easy for Americans to ignore, but now we realize how civil strife, poverty and organized crime in our backyard endanger us. Conditions have become so desperate that people are leaving El Salvador, Guatemala and Honduras as never before."

From 2014 through 2017, according to *The Wall Street Journal*, immigration authorities in the U.S. and Mexico apprehended more than 335,000 migrants from El Salvador alone. Since 2014, the number of U.S. asylum applications from those countries has quadrupled. Thousands are making their way through Mexico toward our southern border, hoping to be admitted.

Many are running for their lives. El Salvador has the highest homicide rate on the planet. "Migrants from all three countries cite violence, forced gang recruitment, and extortion, as well as poverty and lack of opportunity," reports the New York-based Council on Foreign Relations. Vicious criminal gangs live off drug trafficking, kidnapping and extortion, and governments hobbled by corruption can't stop them. About 60 percent of Hondurans and half of Guatemalans live in poverty.

Given the dire circumstances in Central America, stricter border security—without addressing the origins of the problem—is futile. "You have asylum seekers saying I'd rather be in jail in the U.S. than killed in my own country," Maureen Meyer of the Washington Office on Latin America, a human rights organization, told the *Journal*.

The *Chicago Tribune* editorial asserts that "The U.S. should address the migrant crisis at its source. These countries need improved governance, economic growth and public safety. In the early 2000s, Washington largely succeeded with Plan Colombia, which helped end a leftist insurgency and overcome the power of criminal drug cartels in a country once hopelessly out of control. A similar effort would help Central America.

The *Tribune* agrees that if Trump were to slash U.S. aid to Central American countries in revenge for the border problem, "that would be self-defeating. The U.S. should build on the Alliance for Prosperity, which was undertaken with support from President Barack Obama and a Republican-controlled Congress. The program initially allocated $750 million a year in economic and security aid, conditioned on recipient governments making progress on corruption control, policing and human rights. It's a good start, but more may be needed."

The newspaper summarizes, the "U.S. can also provide assistance and counsel on how to improve police agencies and courts, combat corruption, strengthen institutions of civil society, protect property rights and collect taxes. That would require a sustained bipartisan commitment over years if not decades. If it succeeds, it would not only stem the flood of migrants northward; it would pay lasting dividends to the peace and security of the Western Hemisphere.

"The alternative? More chaos, more crime and more caravans. Ignoring Central America, it turns out, isn't an option."

About that border wall; it's a worse idea than you think

When you more closely examine Trump's idea of building a wall across our southern border, you realize it is a phoney, "wedge" issue and not a serious proposal for addressing a real-life problem.

That's probably why, according to the Pew Research Center, people who live less than 350 miles from the border are the least likely to support Trump's wall. In other words, the people supposedly on the front lines of what Trump calls a "crisis" are those least inclined to support the proposed solution to it. The political rhetoric is feeding real fear—among people who live far from the border. But people who live closer to it simply don't identify with that. At one point, Fox News showed video of rioters climbing a wall to make viewers think it was at our southern border—but it wasn't even video shot in our country.

Vox reports "The point of walls is to prevent people from crossing into the U.S. undetected. That's not what most of the families and children who are crossing are doing. They're turning themselves in to the nearest border agent they see on the U.S. side.

"Not all of them, but a large share, are seeking asylum—seeking to live legally in the U.S. That's something they have a legal right to do even if they crossed illegally—and it's something they could do at a port of entry even if there were a wall across the entire border."

If Trump succeeds in building his border wall, it will have a disastrously negative impact on the environment—to say nothing of the hostility it will arouse amongst property owners whose land would have to be seized under "imminent domain" laws. Two-thirds of the property needed for such a wall is privately owned.

Trump's wall would stretch for 1,954 miles from the Gulf of Mexico in Texas to the Pacific Ocean in California, and traverse one of the nation's most diverse landscapes. It would span six separate eco-regions, ranging from desert scrub to forest woodlands to wetland marshes, both freshwater and salt. The wall would bisect 1,506 native animals and plants, including 62 species that are listed as endangered.

A wall would also affect water flows and the patterns of wildfire, exacerbating the risks to people, animals and structures nearby.

There is already evidence for the harm a border wall would do. During the administration of George W. Bush, disastrous flooding occurred in Arizona after 700 miles of fencing was constructed. The barriers acted as dams during rainy season flash floods. In 2008, at the Organ Pipe Cactus National Monument in southwest Arizona, a five-mile-long segment of 15-foot-high wire mesh fence trapped debris flowing through a natural wash during a 90-minute summer thunderstorm, causing water to pool up to seven feet high.

The same storm sent torrents into the city of Nogales, Arizona, a border town 66 miles south of Tucson, causing millions of dollars in property damage. In 2011, another deluge at Organ Pipe knocked over a segment of fence, and in 2014, the twin cities of Nogales (on the U.S. and Mexico sides) flooded again after border barriers got clogged with debris during a heavy storm.

A wall like what Trump wants to build could disconnect a third of 346 native wildlife species from 50 percent or more of their range that lies south of the border. That poses a significant risk to their survival by shrinking and isolating animal populations and limiting their ability to roam for food, water, and mates. Fencing also traps wildlife from escaping fires, floods, or heat waves. Even the pygmy owl is at risk, because when flying, its range is less than five feet off the ground.

The National Geographic Society notes that "Border fencing disrupts seasonal migration, affecting access to water and birthing sites for Peninsular bighorn sheep that roam between California and Mexico. The inability to cross the border has fragmented populations of

Sonoran pronghorn, and diminished the chances of re-establishing colonies of the Mexican gray wolf, jaguars, and ocelots in their range in the United States. Jaguars once roamed the banks of the Rio Grande, but have virtually disappeared from Texas."

Limits on migration, in turn, affects plants. The seeds of mesquite trees germinate best after they have passed through the digestive systems of javelinas and coyotes, according to a report by Defenders of Wildlife.

Proposals under consideration would locate the wall through seven Texas wildlife conservation areas, including the Lower Rio Grande Valley National Wildlife Refuge and Big Bend National Park, prized among national parks as a place so remote it is considered to be one of the best places in the Lower 48 to view the night sky.

In Mission, Texas, the National Butterfly Center, where more than 200 butterfly species live near the banks of the Rio Grande, has been notified that the wall will divide the 100-acre sanctuary, placing almost 70 percent of it on the Mexican side. Plans also call for bisecting a wildlife refuge and state park, placing most of the land on the Mexican side.

Due to an Act of Congress in 1995, in the aftermath of the 9/11 terrorist attacks, the Department of Homeland Security is empowered to waive any laws in the name of national security. So construction of the border wall does not have to meet the requirements of more than 30 of the most sweeping and effective federal environmental laws, such as the Endangered Species Act, National Environmental Policy Act, the Clean Air Act, and the Clean Water Act.

The fears of Americans must be acknowledged

Let's stipulate that no society can remain free with open borders. You can't simply let people walk across a border without having some means of determining—humanely as well as effectively—who they are and why they've come.

As Andrew Sullivan writes in *New York* magazine, "All of it is putting unprecedented strain on liberal democracy in the West itself. The connection between mass migration and the surge in far-right parties in Europe is now indisputable. Without this issue, Donald Trump would not be President. As we can see right now in front of our eyes, elections can turn on this. Which is why Trump is hyping this caravan story to the heavens."

Andrew Sullivan, who concedes that Trump's child separation policy at the border is "hideous and indefensible," nevertheless has given voice to a truism that is often muted on the left. He comments, "We have forgotten in our good intentions that human beings are tribal creatures, that race is a part of tribal identities, and that the racial composition of a country, for good and ill, is something people care about. But you can only ignore human nature for so long until it comes back to bite you.

"It's that force of human nature—often an irrational but still potent force—that has given us the agonies of Brexit and Trump. A revolt against mass nonwhite immigration lies at the heart of both developments."

Sullivan says that "David Frum is right: 'If liberals insist that only fascists will defend borders, then voters will hire fascists to do the job liberals will not do.' And unless the Democrats get a grip on this question, and win back the trust of the voters on it, their chance of regaining the Presidency is minimal. Until one Democratic candidate declares that he or she will end illegal immigration, period, shift legal immigration toward those with skills, invest in the immigration bureaucracy, and enforce the borders strongly but humanely, Trump will continue to own this defining policy issue in 2020."

Andrew Sullivan argues that "At some point, sooner rather than later, we will have to find a point between allowing mass immigration and outflanking white nationalism. Yes, there are racists; but there are also many who are simply uncomfortable and conservative. The only way to stop the populist, white-nationalist right is to separate the former from the latter, and engage the conservatives."

He concludes, "We can and should call out xenophobia and racism, but we also need to accommodate the deep human need for continuity in national and community life. Especially if the alternative only alienates those who are afraid, and empowers the demagogues who exploit them."

❏

A Green New Deal needs to address the issue of immigration and not allow fascist thinking to define it. Polls show older Americans especially are wary of an unregulated influx of foreigners, fearing it could transform our society negatively. But during the 1950s, when President Eisenhower launched the interstate highway system, the U.S. actually recruited foreigners to come here and help build America together. When immigration is not politicized, and has adequate funding and staffing to process applicants for asylum, these migrants can be trained to help build a better, healthier future for all of us.

12.
Our Country in Crisis

N o PRESIDENT OF THE UNITED STATES HAS EVER BEHAVED LIKE Donald Trump—an unprincipled, narcissistic, reckless, mendacious and greedy man of enormous insecurities and grandiosity, who has disrespected American laws, institutions and people. He has unscrupulously enriched himself for decades, and there are ongoing investigations into possible tax and bank fraud. His foreign ties are troubling. As serious thinkers contemplate a Green New Deal to chart a better, brighter future for our country, Trump's dark influence looms large over that effort.

A good measure of the man is the decadent lifestyle he leads.

Covering 20 acres of prime beachfront real estate in Florida, Mar-a-Lago is Trump's 128-room house that was built between 1924 and 1927 for cereal heiress Marjorie Merriweather Post. The lavish rooms are an homage to European opulence, described as "Hispano-Moresque" style, with Italian influence in the form of Florentine frescoes and Venetian arches.

When the estate went on the market for $20 million, Donald Trump offered $15 million. After this offer was rejected, Trump purchased the beach in front of the property and threatened to build a structure to obstruct the ocean view of Mar-a-Lago. He ended up getting Mar-a-Lago for only $5 million, plus an additional $3 million for the antiques and furniture in 1985. He now operates part of the residence as an exclusive club to defray expenses and it has appreciated in value to $100 million. He meets wealthy benefactors there.

He's added $7 million worth of gold leaf and a 20,000-square foot ballroon. A portrait painted by Ralph Wolfe Cowan depicts Trump as a "Sun God" and came to be titled "The Visionary."

It costs taxpayers $3.4 million each time he goes to his Florida estate and $183 million of taxpayer money is required per year to protect Trump Tower in New York City. Based on the number of times he has played golf since taking office, Trump would spend as much as 745 days of his Presidency at a golf course if he wins a second term.

He tweets around the clock—37,799 times by early May 2019. His official White House schedule reveals he does little work, but watches TV incessantly, especially Fox news, which he helps promote.

Mostly he issues impulsive edicts without the merest consultation with his staff, is said to ignore briefing papers, and often changes his mind, vacillating between extremes and sending mixed signals.

We have a dysfunctional President and a country in crisis as a result of it. Our Founding Fathers designed a system of checks and balances which Trump and his enablers have ignored, and which Republicans in Congress have refused to enforce.

Former Treasury Secretary Robert Reich has outlined the situation thusly:

"Donald Trump's goal is, and has always been, division and disunion. It's how he keeps himself the center of attention, fuels his base and ensures that no matter what facts are revealed, his followers will

Curiously Donald Trump repeatedly uses a hand gesture—joining his thumb and index finger together to form a circle—which is also a signal used by white supremacists. Trump adviser Roger Stone even flashed the sign with members of The Proud Boys, a far-right neo-fascist organization. But use of it accords "plausible deniability."

stick by him. But there's another reason Trump aims to divide—and why he pours salt into the nation's deepest wounds over ethnicity, immigration, race and gender. He wants to distract attention from the biggest and most threatening divide of all: the widening imbalance of wealth and power between the vast majority, who have little or none, and a tiny minority at the top who are accumulating just about all. 'Divide and conquer' is one of the oldest strategies in the demagogic playbook: keep the public angry at each other so they don't unite against those who are running off with the goods.

"Over the last four decades, the median wage has barely budged. But the incomes of the richest 0.1 percent have soared by more than 300 percent and the incomes of the top 0.001 percent—the 2,300 richest Americans—they have soared by more than 600 percent. This enormous imbalance is undermining American democracy.

"Martin Gilens of Princeton and Benjamin Page of Northwestern concluded a few years ago that 'The preferences of the average American appear to have only a minuscule, near-zero, statistically non-significant impact upon public policy.' After analyzing 1,799 policy issues that came before Congress, they found that lawmakers only respond to the demands of wealthy individuals and moneyed business interests. No secret here. In fact, Trump campaigned as a populist—exploiting the public's justifiable sense that the game is rigged against them. But he hasn't done anything to fix the system. To the contrary, his divide-and-conquer strategy as President has disguised his efforts to reward his wealthy donors and funnel more wealth and power to those at the top. Trump's tax cuts, his evisceration of labor laws, his filling his cabinet and sub-cabinet with corporate shills, his rollbacks of health, safety, environmental and financial regulations, all have made the super-rich far richer, at the expense of average Americans. Meanwhile, he and his fellow Republicans continue to suppress votes. Senate Republicans have denounced Democratic proposals to increase turnout even calling the idea of making election day a federal holiday 'a power grab' (in the words of Mitch McConnell). Well of course it's a power grab—for the people. Trump and his enablers would rather opponents focus on the ethnic, racial and gender differences he uses to divide and conquer. Don't fall for it. We must be united to take back our democracy."

All of this should illustrate the urgency and imperative for us Americans to wield our influence at the ballot box, and organize at the grassroots level, to embrace a Green New Deal...before it's too late.

13.
Afterword

A S SOMEONE WHO GREW UP AS A MEMBER OF THE BABY BOOM generation, living in a northern climate, I can personally attest to the dramatic ways in which our climate has changed. When I was a kid, we were lucky to have a white Christmas, but now snow is starting to fall as early as mid-September and lasting until May. The winters are also more brutal. So much so that I decided at long last to escape to the South where I find it easier to cope with the heat and humidity than northern wind chills that have been so extreme it has at times felt like it was -70 degrees below zero where I am from.

I live now on a woodland, traversed by a mountain stream. From my patio I can gaze out on a panoply of trees in varying shades of green, backlit by sunshine on most afternoons that makes the leaves sparkle as if festooned by diamonds. I especially like the breezy days when the waving branches create a rustling, soothing sound that is accompanied by the babbling of water flowing over rocks and by a bird song symphony of colorful feathered friends, most of whom I have not yet learned to identify.

But as I was wrapping up work on this book, I was excited one recent afternoon to hear a very strange toy trumpet-like bird call I had never heard before. Suddenly, an exotic-looking bird flew low over my yard, from right to left in my line of vision, and landed on the side of a broken-off, dead tree in my neighbor's yard. My first reaction was that this creature must have escaped from a zoo. It was HUGE. I stood there watching and listening to it's jackhammer-like drilling, just below the broken-off part of the dead tree. Then I quickly snapped some pictures with my camera phone, and when I went indoors and Googled to try to identify it, I was surprised to learn I had just seen a bird that was thought to be extinct for nearly 70 years!

It turns out that what I had observed was an ivory-billed wood-pecker, which is about 20 inches long, with a wing-span of 30 inches.

The bird is shiny blue-black with white markings on its neck and back and extensive white on the trailing edge of both the upper- and underwing (which I had observed when it was sort of sideways in flight). The crest is black along its forward edge, changing abruptly to a bright red on the side and rear. The chin of an ivory-billed wood-pecker is black. When perched with the wings folded, you can see a large patch of white on the lower back, roughly triangular in shape. Among North American woodpeckers, the ivory-billed woodpecker is unique in having a bill whose tip is quite flat-tened laterally, shaped much like a beveled wood chisel. Overall it is a very large and distinc-tive woodpecker with a charismatic, very clean and smooth appearance

There was no mistaking the bird I saw for the more common pileated woodpecker, be-cause the markings are so distinctive.

I was intrigued to learn that back in 2008, the Nature Con-servancy had offered a $50,000 reward to any-one who could provide proof that the bird still existed. The National Geographic Society had even dispatched some ornithologists and a film crew to a southern swampland to try to cap-ture video of this rare woodpecker that someone had spotted. Alas, all they got was some blurry film of the bird in flight—but enough so that they could confirm it was, indeed, an ivory-billed woodpecker. How I wish I had shot video of "my" bird!

I learned from Wikipedia that "Habitat destruction and, to a lesser extent, hunting have decimated populations so thoroughly that the species is very probably extinct, though sporadic reports of sightings have continued into the 21st century. The ivory-billed wood-pecker, dubbed the 'holy grail bird' due to its appearance and behavior, is the subject of many rediscovery efforts and much speculation." I also learned it is sometimes referred to as the "Good God Bird," based on the exclamations of surprised onlookers, and that utterance is pretty close to what I said when I saw it. Another nickname is "Elvis In Feathers."

Wikipedia states that "The species is listed as critically endangered by the International Union for Conservation of Nature. The American Birding Association lists the ivory-billed woodpecker as a class 6 species, a category it defines as 'definitely or probably extinct.'

"Reports of at least one male ivory-billed woodpecker in Arkansas in 2004 were investigated and subsequently published in April 2005 by a team led by the Cornell Lab of Ornithology. No definitive confirmation of those reports emerged, despite intensive searching over five years following the initial sightings."

Wikipedia says that "Despite published reports from Arkansas, Florida, and Louisiana, and sporadic reports elsewhere in the historic range of the species since the 1940s, no universally accepted evidence exists for the continued existence of the ivory-billed woodpecker."

But I know I saw one.

Prior to the Civil War, much of the southern United States was covered in vast and continuous primeval hardwood forests that were suitable as habitat for these colorful birds. But after the war, the timber industry deforested millions of acres in the South. Heavy logging activity exacerbated by hunting by collectors devastated the population of ivory-billed woodpeckers in the late 19th century.

This unfortunate occurrence also affected other birds, who were killed because of the fashion of the day: many women wore hats adorned with colorful feathers. This particular species of woodpecker was generally considered extinct in the 1920s when a pair turned up in Florida, only to be shot for specimens.

By 1938, only 20 woodpeckers were estimated to remain in the wild, some six to eight of which were in the old-growth forest called the Singer Tract, owned by the Singer Sewing Company in Madison Parish in northeastern Louisiana, where logging rights were held by the Chicago Mill and Lumber Company. The company brushed aside pleas from four Southern Governors and the National Audubon Society with extensive field studies and a report that the tract be publicly purchased and set aside as a reserve. By 1944, the last known ivory-billed woodpecker, a female, was gone from the cut-over tract.

Having just about completed the manuscript for this book on the topic of our needing a Green New Deal—due to the despoilation of our environment and the concomitant harm to our health, economy and national security that has occurred in recent years—I found it ironic that I personally observed such a rare bird in my own backyard.

I felt like this experience was further proof that Nature is still struggling to preserve life as it once was, but needs an immediate assist.

This underscored for me the importance of our not allowing cynical politicians to deceive us into believing that letting industry ravenously exploit our natural resources is for the common good.

Equally absurd, to my way of thinking, is the notion propounded by some so-called Christians who say we're living in what the Bible defines as the "end times" anyway, and since we don't know when the Second Coming will occur, we needn't worry about the future, and should just go ahead and exploit what we have now. I don't profess to be a Biblical scholar, but I can find no Scripture which gives humans encouragement to abuse the earth.

God commands us to keep and care for the earth but God's edict to have dominion over the earth doesn't mean to have complete domination and exploitation of it. Genesis 2:15 says "The Lord God took the man and put him in the Garden of Eden to work it and keep it."

Jack Wellman, Pastor of the Mulvane Brethren church in Mulvane, Kansas explains it thusly: "God gave mankind a command and told him that he must tend or keep the garden. The Hebrew word for 'tend' or some translations say 'keep' it is 'shamar' and it means more than just keep it neat and tidy. The Hebrew word means 'to guard' or 'to watch and protect.' The other Hebrew word in this verse that's very important is the word 'work' or as some translations more accurately say 'to cultivate' and is from the Hebrew word '`abad' meaning 'to serve.' So Genesis 2:15 would better read as: 'The Lord God took the man and put him in the garden of Eden to serve it and to guard and protect it.' That means that we are stewards of the earth and the Master will require of us an account on how we've been stewards of what He has given us. So far, it's not been good stewardship to say the least...

"We continue to rape the land, denude the forests, strip-mine the hills, burn through the resources, and gut the earth. This is the only place that we have and if we chose to abuse it, it will be a choice we will live to regret."

Numbers 35:33 is unequivocal: "You shall not pollute the land in which you live..."

We are admonished to study the earth and the environment. The Bible says that knowledge of God and his works can be found in understanding plants and animals and the delicate web of life:

"But ask the animals, and they will teach you, or the birds of the air, and they will tell you; or speak to the earth, and it will teach you, or let the fish of the sea inform you. Which of all these does not know that the hand of the Lord has done this? In His hand is the life of every creature and the breath of all mankind." (Job 12:7-10)

God commands humans to be stewards of the environment. Isaiah and Jeremiah prophesy about the dire consequences that occur when man disobeys God and fails to take care of the earth:

"I brought you into a fertile land to eat its fruit and rich produce. But you came and defiled my land and made my inheritance detestable." (Jeremiah 2:7)

We've come full circle

In going back over my book a historical pattern emerges. The first white immigrants to our land tended to exploit what they perceived to be a limitless abundance of resources, with no thought given to tomorrow. Species became extinct. Forests were destroyed. Polluted water made people sick. Air was befouled. Soil was exhausted and blew away, as in the Dust Bowl years. But they acted out of ignorance.

Finally, there was a great awakening, and some of our Presidents took the lead through executive or legislative action—in league with motivated citizens—to protect our natural assets not only for themselves but for future generations.

But with the arrival of Ronald Reagan in the White House, the clock was turned back. He and the three Republican Presidents who have followed him—the two Bushes and now, Donald Trump—have abandoned the bipartisan efforts of past Presidents to protect our natural resources. These four Presidents didn't do their dirty deeds by themselves; they had a lot of help from members of Congress from both parties, as well as big corporate interests. And while Barack Obama can be credited with doing far more to help the environment than the four Presidents to whom I just referred, his mishandling of the BP oil spill in the Gulf of Mexico tarnishes his record considerably.

The thing about it is, we are far more technologically advanced now than the earliest settlers were. We are not as ignorant of the repercussions of our actions.

NASA, for example, maintains a fleet of earth science spacecraft and instruments in orbit studying all aspects of the earth system (oceans, land, atmosphere, biosphere, cryosphere), with more planned for launch in the next few years.

The agency's research encompasses solar activity, sea level rise, the temperature of the atmosphere and the oceans, the state of the ozone layer, air pollution, and changes in sea ice and land ice. NASA scientists regularly appear in the mainstream press as climate experts to explain the ominous changes to our earth that are occurring.

The new United Nations report, based on the input of 500 scientists, warns of biodiversity loss, which they have quantified.

Yet House Republicans are wielding the Green New Deal, and by extension climate action, as a campaign cudgel as a key part of their 2020 election strategy to cast freshmen Democrats as radicals who are waging an "assault on American democracy."

They do so with the prospect of the near- and long-term loss of millennials and Generation Z voters, a growing slice of the electorate that wants federal climate change action to a greater degree than their elders. A recent Harvard Kennedy School Institute of Politics survey found 74 percent of likely general election voters under 30 disapprove of Trump's climate change stance. Their turnout ratio increased by 16 points in 2018, and there's reason to believe that uptick will continue in 2020. Another poll showed climate change is "top of mind" for voters under 40, whether they live in a coastal area or not.

But extreme weather events are capturing the attention of people nationwide, regardless of age or party affiliation. Areas that never used to flood have found themselves under water for weeks. Hurricanes are more frequent and intense. Droughts are lasting longer and are more severe. Even if they aren't paying close attention to the headlines, people are viscerally becoming aware of something happening.

We need to reassert our collective influence to stop those who would perpetuate the status quo before they destroy our planet. Therefore, we need a blueprint, a plan of action, that can help us to coordinate nationwide efforts to make significant changes in how we live, in order to become less dependent on fossil fuels, retool for energy efficiency, and reinvest in American infrastructure to take advantage of new technologies and create good paying jobs that will restore some balance and address the societal scourage of extreme income disparity.

The **onus** is on us to move forward. In other words, our future is *on us*.

Selected Bibliography

Abbey, Edward. *Desert Solitaire: A Season in the Wilderness.* 1967, 1990.

Arms, Myron. *Riddle of Ice: A Scientific Adventure into the Arctic.* 1999

Bjornerud, Marcia. *Timefulness: How Thinking Like a Geologist Can Help Save the World.* 2018.

Bormann, F. Herbert and Stephen Kellert (eds.). *Ecology, Economics, Ethics: The Broken Circle.* 1993.

Bridle, James. *New Dark Age: Technology and the End of the Future.* 2018.

Brown, Kenneth. *Four Corners: History, Land and People of the Desert Southwest.* 1996.

Burns, Loree Griffen. *Tracking Trash: Flotsam, Jetsam, and the Science of Ocean Motion.* 2007.

Carter, Jimmy. *An Outdoor Journal: Adventures and Reflections.* 1988.

Cohen, Joel. *How Many People Can the Earth Support?* 1996.

Carson, Rachel. *Silent Spring.* 2002 edition.

Colbert, Elizabeth. *The Sixth Extinction: An Unnatural History.* 2014.

Cone, Marla. *Silent Snow: The Slow Poisoning of the Arctic.* 2006.

Cronon, William. *Changes in the Land: Indians, Colonists, and the Ecology of New England.* 1983, 2003.

Devall, Bill and George Sessions. *Deep Ecology: Living as if Nature Mattered.* 1985, 2001.

Dillard, Annie. *Pilgrim at Tinker Creek.* 1975, 2007.

Ehrlich, Gretel. *The Solace of Open Spaces.* 1986, 1992.

Ehrlich, Paul. *The Stork and the Plow: The Equity Answer to the Human Dilemna.* 1997.

Forsyth, Adrian and Ken Miyata. *Tropical Nature: Life and Death in the Rain Forests of Central and South America.* 1987.

Gilbert, Sara. *The Imperfect Environmentalist: A Practical Guide to Clearing Your Body, Detoxing Your Home, and Saving the Earth (Without Losing Your Mind).* 2013.

Goodall, Jeff. *The Water Will Come: Rising Seas, Sinking Cities, and the Remaking of the Civilized World.* 2017.

Graedel, Thomas, and Paul Crutzen. *Atmosphere, Climate and Change.* 1997.

Hardin, Garrett. "The Tragedy of the Commons." *Science* 162 (December 13, 1968): 1243-1248.

Henson, Robert. *The Rough Guide to Climate Change.* 2006.

Hertsgaard, Mark. *Earth Odyssey: Around the World in Search of Our Environmental Future.* 1998, 2000.

Houle, Marcy. *The Prairie Keepers: Secret of the Grasslands.* 1996.

Leakey, Richard and Roger Lewin. *The Sixth Extinction: Patterns of Life and the Future of Humankind.* 1996.

Leonard, Annie. *The Story of Stuff: The Impact of Overconsumption on the Planet, Our Communities, and Our Health—and How We Can Make It Better.* 2010.

Leopold, Aldo. *A Sand County Almanac.* 1968, 2001.

Lopez, Barry. *Of Wolves and Men.* 1979, 2004.

Mann, Charles C. *1491: New Revelations of the Americas before Columbus.* 2005.

Muir, John. *My First Summer in the Sierra.* 1911, 2004.

Nash, Roderick. *Wilderness and the American Mind.* 2001.

National Audubon Society, *Audubon Society Field Guide.* 1981.

Preston, Richard. *The Wild Trees: A Story of Passion and Daring.* 2007.

Quammen, David. *Song of the Dodo: Island Biogeography in an Age of Extinctions.* 1997.

Reisner, Marc. *Cadillac Desert: The American West and Its Disappearing Water.* 1986, 1993.

Rich, Frederic C. *Getting to Green: Saving Nature: A Bipartisan Solution.* 2016.

Rush, Elizabeth. *Rising: Dispatches from the New American Shore.* 2018.

Scarce, Rick. *Eco-warriors: Understanding the Radical Environmental Movement.* 1990, 2005.

Terborgh, John. *Where Have all the Birds Gone?* 1990.

Thoreau, Henry David. *Walden.* 2012 edition.

Tickle, Josh. *Kiss the Ground: How the Food You Eat Can Reverse Climate Change, Heal Your Body & Ultimately Save Our World.* 2017.

Turco, Richard. *Earth Under Siege: From Air Pollution to Global Change.* 1996, 2002.

Wessels, Tom. *Reading the Forested Landscape: A Natural History of New England.* 1999.

Wilkinson, Charles. *Crossing the Next Meridian : Land, Water, and the Future of the West.* 1993.

Willis, Delta. *The Sand Dollar and the Slide Rule: Drawing Blueprints from Nature.* 1996.

Wilson, Edward O. *The Diversity of Life.* 1999, 2003.

https://www.wilderness.org/articles/article/20-popular-american-wilderness-areas

https://wonderopolis.org/wonder/are-there-any-wildernesses-left-in-america

https://www.uua.org/action/statements/threat-global-warmingclimate-change

https://www.wilderness.org/articles/article/trump-aims-attack-national-monuments-27-risk

https://usaforests.org

https://www.theguardian.com/environment/2018/dec/05/american-wilderness-trump-energy-threat

https://www.splcenter.org/hatewatch/2018/09/18/ok-sign-white-power-symbol-or-just-right-wing-troll

https://www.theatlantic.com/politics/archive/2018/08/unite-the-right-charlottesville-trump/567263

https://www.nature.org/en-us/what-we-do/our-priorities/tackle-climate-change/climate-change-stories

https://www.nationalgeographic.com/environment/2019/01/how-trump-us-mexico-border-wall-could-impact-environment-wildlife-water

https://www.climaterealityproject.org/blog/how-climate-change-affecting-florida

https://www.uua.org/action/statements/threat-global-warmingclimate-change

https://www.smithsonianmag.com/smithsonian-institution/the-real-birth-of-american-democracy-83232825

https://www.theguardian.com/environment/2018/dec/17/climate-change-activists-vow-step-up-protests-around-world

https://www.wilderness.org/articles/article/20-popular-american-wilderness-areas

https://www.quora.com/Why-does-socialism-have-such-a-bad-name-in-the-USA-Is-it-only-because-it-is-often-confused-with-communism-or-are-there-other-reasons

https://www.uua.org/action/statements/threat-global-warmingclimate-change

https://www.jurist.org/commentary/2009/10/our-socialist-founding-fathers

https://www.theguardian.com/environment/20

https://grist.org/article/whats-the-green-new-deal-the-surprising-origins-behind-a-progressive-rallying-cry

http://nymag.com/intelligencer/2017/05/republican-sick-people-dont-deserve-affordable-care.html?fbclid=IwAR1hZAkBlhjt6njG0EAz2lFNU6JF7tlvvviGtn848n6mujABRnwlDZVi4o

https://www.aft.org/resolution/crime-against-humanity

https://www.nature.org/en-us/what-we-do/our-priorities/tackle-climate-change/climate-change-stories

https://www.thedailybeast.com/inside-fdrs-wild-obsession-and-jeffersons-passion-for-the-land

http://time.com/5524723/green-new-deal-history

http://trn.trains.com/railroads/railroad-history/2011/01/civil-war-rails

https://www.reference.com/history/were-effects-transcontinental-railroad-3f1b91e-5b38101d8?qo=contentSimilarQuestions

https://www.nps.gov/nr/travel/presidents/hoover_camp_rapidan.html

https://www.sciencehistory.org/distillations/magazine/richard-nixon-and-the-rise-of-american-environmentalism

http://ak.audubon.org/conservation/arctic-national-wildlife-refuge

https://www.kcet.org/redefine/green-urbanism-remembering-jfks-major-environmental-achievements

https://www.history.com/topics/great-depression/dust-bowl

https://transcription.si.edu/project/7817

https://opinionator.blogs.nytimes.com/2014/11/15/the-civil-wars-environmental-impact

https://www.thebalance.com/what-was-the-dust-bowl-causes-and-effects-3305689

https://shvpl.info/imagecgkl-camp-david-house-inside.htm

https://stampaday.wordpress.com/2018/01/30/president-franklin-delano-roosevelt-philatelist

https://www.whitehousehistory.org/the-fireside-

chats-roosevelts-radio-talks
https://cleantechnica.com/2014/04/27/jimmy-carter-environmental-advocate-inspiring-leader
https://www.terrapass.com/us-presidents-environmental-legacies
https://www.nps.gov/grsm/learn/historyculture/stories.htm
https://www.thebalance.com/what-was-the-dust-bowl-causes-and-effects-3305689
https://www.nytimes.com/2016/08/16/business/dealbook/the-risks-of-unfettered-capitalism.html
https://blog.nwf.org/2013/12/restoring-the-everglades-restoring-the-gulf
https://wikispooks.com/wiki/Zapata_Petroleum
https://whowhatwhy.org/2013/10/14/bush-and-the-jfk-hit-part-5-the-mysterious-mr-demohrenschildt
http://paulkangas.tripod.com/ghwbushlbjkilledpresidentjohnkennedy
http://www.ens-newswire.com/ens/jan2001/2001-01-19-06.html
https://www.refugeassociation.org/advocacy/refuge-issues/arctic
https://www.classicdriver.com/en/article/classic-life/presidential-style-jfk-factor
https://www.politico.eu/article/attacks-will-be-spectacular-cia-war-on-terror-bush-bin-laden
https://history.nasa.gov/moondec.html
https://www.usnews.com/news/the-report/articles/2015/08/28/hurricane-katrina-was-the-beginning-of-the-end-for-george-w-bush
https://www.space.com/17547-jfk-moon-speech-50years-anniversary.html
https://www.nasa.gov/50th/50th_magazine/benefits.html
https://www.theguardian.com/politics/2009/jan/16/greenpolitics-georgebush
https://www.csmonitor.com/Environment/Bright-Green/2008/0527/smearing-rachel-carson
https://socialistworker.org/2017/01/06/the-obama-years-of-squandered-hopes
https://theweek.com/articles/479867/federal-reserves-breathtaking-77-trillion-bank-bailout
https://www.nytimes.com/2009/01/18/business/18bank.html
https://www.lwcfcoalition.com/about-lwcf
https://www.dw.com/en/opinion-obamas-environmental-legacy-of-missed-opportunities/a-36863223
http://www.trumptwitterarchive.com/archive
https://www.visitcalifornia.com/in/attraction/richard-nixon-presidential-library-and-museum
https://www.nytimes.com/roomfordebate/2012/06/13/did-any-good-come-of-watergate/nixon-

had-some-successes-before-his-disgrace
https://www.cnn.com/2016/04/21/opinions/when-americans-fought-over-gasoline-opinion-jacobs/index.html
https://thinkprogress.org/a-graphical-look-at-presidents-environmental-records-f232f07005d0
https://www.thenation.com/article/running-on-empty
https://www.nytimes.com/interactive/2017/10/05/climate/trump-environment-rules-reversed.html
https://millercenter.org/president/carter/life-after-the-presidency
https://dangerousminds.net/comments/reagan_finally_does_what_hes_been_waiting_years_to_do
https://www.politico.com/story/2019/01/26/puerto-rico-hurricane-aid-1125530
http://www.underworldtales.com/ghosts-of-911
https://www.biography.com/news/911-anniversary-facts
https://www.sierraclub.org/take-action
https://grist.org/climate-energy/top-10-reasons-why-fracking-for-dirty-oil-in-california-is-a-stupid-idea
https://iowastartingline.com/2019/03/03/the-green-new-deal-isnt-new-obamas-already-proven-it-can-work
https://friendsofanimals.org/news/obama-creates-first-marine-monument-in-atlantic-off-limits-to-fishing-drilling
https://www.dw.com/en/opinion-obamas-environmental-legacy-of-missed-opportunities/a-36863223
https://www.cbsnews.com/news/stuck-inside-tour-a-few-national-parks-with-president-obama
https://blackamericaweb.com/2016/11/01/obamas-reveal-private-residence-in-white-house
https://environmentamerica.org/blogs/environment-america-blog/ame/president-obama%E2%80%99s-renewable-energy-legacy
https://www.vox.com/policy-and-politics/2018/2/2/16956014/nunes-memo-carter-page
https://www.eia.gov/tools/faqs/faq.php?id=727&t=6
https://www.rt.com/business/448706-russia-record-coal-exports
https://www.abc.net.au/news/2018-06-11/donald-trump-russia-carter-page-speaks-to-four-corners/9846290
https://www.npr.org/2016/08/06/488876597/how-the-trump-campaign-weakened-the-republican-platform-on-aid-to-ukraine

https://www.americangeosciences.org/critical-issues/faq/how-much-coal-does-us-export-and-import
https://www.wbur.org/cognoscenti/2018/08/22/trump-epa-coal-pollution-fred-hewett
http://www.environmentalintegrity.org/trumpwatch-epa/whos-running-trumps-epa
http://foreverlogical.com/epa-environment
https://thinkprogress.org/interior-scandals-bernhardt-investigation-bears-ears-ethics-violations-b095ba71cea8
https://www.politico.com/story/2018/12/21/ryan-zinke-montana-interior-career-1070944
https://ktla.com/2018/06/28/schwarzenegger-mocks-trump-on-coal-industry-asks-if-hell-bring-back-blockbuster-next
https://timeline.com/la-smog-pollution-4ca4bc0cc95d
https://thinkprogress.org/interior-scandals-bernhardt-investigation-bears-ears-ethics-violations-b095ba71cea8
https://www.nationalreview.com/2019/04/trump-making-speeches-fun-again-at-nra-meeting
https://www.huffpost.com/entry/white-evangelicals-trump-morality_n_5cc20d-6de4b031dc07efb940
https://grist.org/article/whats-the-green-new-deal-the-surprising-origins-behind-a-progressive-rallying-cry
https://www.nationalparkstraveler.org/2018/08/traveler-special-report-maintenance-backlog-crippling-national-park-roads-and-bridges
https://www.nature.org/en-us/what-we-do/our-priorities/tackle-climate-change/climate-change-stories
http://nymag.com/intelligencer/2017/05/republican-sick-people-dont-deserve-affordable-care.html?fbclid=IwAR1hZAkBlhjt6njG0EAz2lFNU6JF7tlvvviGtn848n6mujABRnwlDZVi4o
https://theculturetrip.com/north-america/usa/texas/articles/17-stunning-photos-of-texas-big-bend-national-park
https://www.usatoday.com/story/news/politics/2013/10/20/cheney-head--and-heart-strong/310704
https://www.theatlantic.com/politics/archive/2011/08/remembering-why-americans-loathe-dick-cheney/244306
http://appvoices.org/2014/12/19/its-still-happening-2
https://www.theguardian.com/us-news/2015/oct/15/organ-pipe-national-monument-migrants-mexico
https://www.expressnews.com/news/local/article/On-watch-for-smugglers-drug-runners-in-the-vast-12416884.php

https://earthjustice.org/features/campaigns/what-is-mountaintop-removal-mining
http://neverendingplaylist.com/billy-andrusco#&song=misty-rain-from-the-album-hollywood-hits&pos=4
https://www.hcn.org/articles/opinion-president-trump-is-sacrificing-our-public-lands-legacy
https://www.cnbc.com/2019/02/16/keystone-xl-suffers-another-setback-judge-blocks-work-on-pipeline.html
https://www.stuartwilde.com/2013/04/the-horrendous-evil-of-dick-cheney-uncovered
https://www.usatoday.com/story/news/nation/2019/01/01/free-all-national-parks-overrun-garbage-other-bad-behavior/2456757002
https://constitutioncenter.org/blog/the-lost-presidential-speech-made-at-the-gettysburg-address-anniversary
https://www.nps.gov/articles/11-ways-national-parks-influenced-world-war-i.htm
https://www.theguardian.com/environment/2019/jan/25/death-valley-national-park-damage-offroaders-shutdown
https://www.mnn.com/earth-matters/wilderness-resources/photos/most-endangered-trees-america/maple-leaf-oak-quercus-acerifolia#top-desktop
https://www.uua.org/action/statements/threat-global-warmingclimate-change
https://pastdaily.com/2016/09/02/fdr-great-smoky-mountains-sept-2-1940
https://www.shorpy.com/node/7772
https://www.taxpayer.net/article/forest-service-loses-88-million
https://www.times.org/forest-liquidation-sale
http://stopthinningforests.org/further-resources2.html
https://www.saveamericasforests.org/resources/Destruction.htm
https://mountainjournal.org/morale-plummets-in-forest-service
https://www.americamagazine.org/faith/2017/1
https://fornology.blogspot.com/2018/04/when-white-pine-was-green-gold-in.html
https://mountainjournal.org/morale-plummets-in-forest-service
https://www.pressdemocrat.com/news/2177821-181/vow-by-new-company-to
https://www.times-standard.com/2019/03/31/friends-of-the-eel-river-offers-hikes-along-great-redwood-trails-future-home
https://www.getredwood.com
https://www.greenpeace.org/usa/victories/kleercut-kimberly-clark-commits-to-end-deforestation
https://www.archives.gov/research/ansel-adams

https://www.washingtonpost.com/graphics/2019/national/gone-in-a-generation/forest-climate-change.html?utm_term=.db14f1ae4bde
https://www.americanforests.org/blog/
https://www.wilderness.org/articles/article/11-americas-greatest-national-forests
http://stopthinningforests.org/further-resources2.html
https://www.nytimes.com/2019/03/13/opinion/northern-rockies-carole-king.html
https://psmag.com/environment/wildfire-and-fury
https://www.theguardian.com/us-news/2018/nov/13/donald-trump-wildfires-science-forest-management
https://www.pressdemocrat.com/news/2177821-181/vow-by-new-company-to?sba=AAS
https://psmag.com/environment/wildfire-and-fury
https://www.factcheck.org/2018/11/trump-repeatedly-errs-on-california-wildfires/
https://www.wood-database.com/wood-articles/ten-best-woods-youve-never-heard/the wood database
https://www.wood-database.com/wood-articles/restricted-and-endangered-wood-species
https://www.5280.com/2018/01/danger-in-the-forest
https://carolinapublicpress.org/28444/interest-group-emails-compete-to-influence-nc-national-forests-future
https://www.buzzfeed.com/whitneyjefferson/coral-bleaching-and-why-you-should-care
https://www.enca.com/life/large-vessels-are-fishing-55-percent-of-worlds-oceans
https://traveltips.usatoday.com/places-snorkeling-jupiter-florida-111395.html
https://www.bbc.com/news/world-europe-45209587
https://www.news.com.au/technology/environment/climate-change/climate-change-could-cause-oceans-to-turn-green-by-2100-study-warns/news-story/675622fbecfa3262ec85411fe62a259b
https://oceana.org
https://www.ourlaststraw.org/about?gclid=CjwKCAjw8LTmBRBCEiwAbhh-6Lj__NKDEqcPNywCewEW2fleCReQq_4ynCSYNnnXx9fl5eQ_tAeWWxoCwrsQAvD_BwE
https://www.fiveaa.com.au/news/Expert-Says-Climate-Change-Could-Turn-Earth-s-Oceans-Green
https://www.peoplesworld.org/article/oceans-in-trouble-17422
https://health2016.globalchange.gov/temperature-related-death-and-illness
https://19january2017snapshot.epa.gov/climate-impacts/climate-impacts-human-health_.html
https://www.aft.org/resolution/crime-against-humanity
https://www.healthaffairs.org/do/10.1377/hblog20181218.278288/full
https://www.healthaffairs.org/do/10.1377/hblog20181218.278288/full/
https://voidlive.com/1-9-0-future-st-augustine-squares-off-sea-level-rise
https://www.ucsusa.org/global_warming/science_and_impacts/impacts/climate-change-adaptation-st-augustine-florida-heritage
https://www.mcclatchydc.com/news/politics-government/white-house/article228724179.html
https://www.vox.com/2018/6/18/17475292/family-separation-border-immigration-policy-trump
http://nymag.com/intelligencer/2018/10/democrats-cant-keep-dodging-immigration-as-a-real-issue.html
https://www.tampabay.com/news/politics/national/Erosion-of-immigrant-protections-began-with-Trump-inaugural_169232939
https://www.nytimes.com/interactive/2019/01/24/us/migrants-border-immigration-court.html
https://www.businessinsider.com/melania-trump-immigration-lawyer-slams-trump-chain-migration-attacks-2018-8
https://www.aft.org/resolution/crime-against-humanity
https://www.chicagotribune.com/news/opinion/editorials/ct-edit-central-america-migrants-caravan-aid-20181105-story.html
https://www.counterpunch.org/2019/01/21/central-america-needs-a-marshall-plan
https://www.nationalgeographic.com/environment/2019/01/how-trump-us-mexico-border-wall-could-impact-environment-wildlife-water
https://thegolfnewsnet.com/golfnewsnetteam/2019/05/05/how-many-times-president-donald-trump-played-golf-in-office-103836

Made in the USA
San Bernardino, CA
15 May 2020

71698513R00168